MUSINGS

ON

LEADERSHIP

It's for Everyone and Everywhere in Life

SgtMaj William J. Singleton, USMCR (ret.)

*Whether you're a Commander, CEO, Pastor, Teacher,
Laborer, Artist, Father, Mother, Husband or Wife—
YES, THIS BOOK IS FOR YOU!*

1

Musings on Leadership
Copyright © 2022 W. J. Singleton

All Scripture references are taken from the ESV® Bible (The Holy Bible, English Standard Version®). Copyright © 2001 by Crossway, a publishing ministry of Good News Publishers. Used by permission. All rights reserved.

ISBN: 979-8-9866850-8-3

Cover design by W. J. Singleton

Printed in the United States of America

Dedication

This book is dedicated to four groups of people. First and foremost, I dedicate this to God and His son, Jesus Christ, and the Holy Spirit. Surely His goodness and mercy have followed me all the days of my life till this point, and I pray that will be the case until my demise. All of my achievements and blessings in life are His alone and are merely seen through me by His grace.

Second, I dedicate this book to my beloved wife and the three wonderful children she has given me. I cannot capture in words the love I have for her and for them. They truly are what make my life worth living.

Third, I dedicate this book to the rest of my family. They have kept the faithful bond of family by encouraging me up to this point, and I believe they will do so until my race is finished.

Lastly, I dedicate this book to my friends and the all the awesome men and women I have had the honor and privilege to serve with, suffer with, and celebrate with throughout my life. I have learned from you, stood in awe of you, been overwhelmingly proud to be counted as one of you, and had my heart filled by knowing you. Fair winds and following seas to you all—past, present, and future.

4

Contents

Foreword

If you're reading this that means you got past the cover and any doubts you may have had as to whether this book would be meaningful or enriching to you. I'm glad that you did, because I believe that it will be. The cover merely shows one part of my life where I've experienced leadership. However, there have been many more besides just the military. And I dare say at this point, much more significant ones. Leadership is a true need for everyone in their lives and all aspects of their lives. It's so important that God established it as a system of man at the very beginning of creation. He gave Adam dominion over all the Earth and its creatures. Hmmm, sounds like a leadership role to me. He then put Adam to sleep and from him created a helpmate called Eve. Hmmm, the term helpmate seems to indicate a leader/lead relationship. That doesn't even begin to speak of the numerous great biblical leaders God established and guided in executing his will. Culminating with the greatest one to walk the Earth. The one who would lead us to salvation, Jesus Christ. One of Jesus' primary missions was to establish the church. Hmmm, the church sounds like an organization made up of people. Jesus then charged 12 apostles with going out, spreading the gospel, and growing the body of the church. Well, that sounds like a leader putting other leaders in place.

It wasn't until I matured in my leadership experience, that I realized that leadership is applicable to and should be occurring, everywhere in everyone's life. Whether a person is the subject of someone else's leadership, or a person is the one doing the leading, leadership is a necessity to human existence. I guarantee that you have a leadership role somewhere in life. As a matter of fact, I can guarantee that you have multiple leadership roles in your life, whether you're choosing to fulfill the duties of those roles or not. The point is, no matter who you are, or what you do in life, the thoughts on leadership covered in this book are pertinent to you. It is my sincere hope that they contribute not only to who you are as a leader now, but your continued growth and future success as one as well.

INTRODUCTION

I hope I have been a leader, but one never knows that for sure as it is something that is determined by those that are led.

Writing this book took me through a refresher course of the leadership lessons, skills, and insights I have gained throughout my life to date. Forty-plus combined years of leadership training and real-life practice in the U.S. Marine Corps and law enforcement has exposed me to a plethora of leadership styles, techniques, and tools—some utilized by me, some not. True leaders—or aspiring ones—know there is always room for improvement and are constantly looking to better themselves. At this stage of my life, I am relatively familiar with who I am as a person and how my core values and communication style affect who I am as a leader. However, a person can only *hope* to lead others; it is the follower who determines whether a person is a leader or not. Since being a leader is not a status or position you obtain and then have forever, a leader must constantly work at being deserving of their title and role. At one of the most impactful leadership presentations I attended, the instructor said that all leadership should be servant-based leadership. Followers deserve and have a right to the best leader they can get. A leader who strives to obtain new skills and is constantly working on self-improvement offers the greatest service to their followers. Another important aspect of leadership is that a person is rarely a leader in every aspect of their life. To be a good leader you must first be a good follower. I have been a follower as often as I have been a leader, and I take as much pride in being an exemplary follower as I do being an exemplary leader. I've been a person who has had supervisors and managers in charge of me, and I've been a person who has been led by leaders. Leadership is indeed determined by those who are led. But just because you're in charge of somebody does not mean you are leading them.

During my time in the Marines, I have had the pleasure to serve as a Senior Enlisted Advisor (SEA) for several units. While serving as a SEA for these units, I had the experience of forming necessary leadership relationships with several different commanding officers (CO). The CO/SEA relationship is a special one—and a uniquely military one too. Leadership of a unit or any organization cannot be done alone; one person may be at the head. But the success or failure of leading an organization is almost never done by a single person. One thing I began doing when I became a SEA was to have a sit-down with my new CO as soon as our professional relationship started so that we could get a good feel for and understanding of each other. As a result of their respect for the experience and knowledge of SEAs, COs often give their SEAs wide latitude in the performance of their duties. Some COs and even some SEAs believe that wide latitude equates to totally independent operation. During the sit-down, I made it a point to not only have the CO explain what they expected of me; I also explained what I expected of them. One thing I found important to express to them was that I was a member of their unit just like every other Marine; therefore I needed and expected them to provide me outstanding leadership as well. This is because all people—even those who are highly competent and committed—need direction, support, and encouragement from time to time. Even people who hold senior-level positions require leadership, though their leadership needs are noticeably different than those who work under them.

You will notice that some concepts throughout this book are repeated. I find this technique important because leadership, at its core, is uncomplicated, yet is comprised of many interrelated concepts.

CHAPTER ONE: Sound Principles

Leadership: I can't tell you what it is, but I know it when I see it.

Well-known writers and famous people the world over have written many works on leadership—what it is, what it looks like, and how to become one. I propose that no one can tell you precisely how to be a leader, or exactly what leadership is, or even how to do it. There are most certainly some quantifiable and practical elements to it, but leadership is something that also falls into the nebulous realm of human psychology and sociology. Only humans can *lead* humans. Only humans can truly be *followers* of other humans. And leadership requires human followers' cooperation in order for leadership to happen. Because human beings comprise the two critical elements in leadership (the leader and the led)—and humans do not operate on a purely rational basis—leadership is more of an art than a science. It is largely based upon intangible things, such as concepts, intuition, insight, and coup d'oeil (the ability to instantly grasp a situation).

So if no one can tell, teach, or explain leadership, then why read any further? Well, while the thing itself cannot be implicitly imparted, different elements that contribute to the essence of leadership can be related, such as the principles and traits that contribute to a person being a good leader.

Yes, some of you will recognize the following as a take on the Marine Corps' version of leadership principles and traits. Well, I'm a Marine, so what do you expect? All kidding aside, besides the fact that I'm a Marine, in looking back over my years of experiencing leadership, I can find no better foundational basis of what contributes to a person being a good leader than the eleven principles and fourteen traits I feature in the next two chapters. At the very least, they are no worse than any others you might have been taught. I've read and studied leadership and been privy to much discussion on the topic by many who were and are considered to be accomplished or well-versed in the area. Almost without fail, I have been able to correlate all of what I've read and heard on the topic to one or more of the following principles and traits. What I endeavor to accomplish is to provide more of the "why" of them rather than the "what." In one of the most profound books that I have read, author Simon Sinek encourages use all to start with *why*.[1] It's unfortunate that as society advances, the *why* of things is less known or answered. It's unfortunate because *why* tells the purpose and purpose is what provides impetus to man's actions. Statistically, when a person understands something intellectually, they are more likely to retain the information. If a person retains something and it makes sense, they are more likely to internalize it. If they internalize it, they are more likely to recognize and apply what they have retained to various life situations as appropriate.

Leadership Principles

Set the example.

I personally believe this is rightfully the first one on the list of principles for two reasons. First, it is simultaneously the easiest and the hardest of them to do. Easiest because it only requires the leader's participation and cooperation; hardest because it requires the leader's concerted will, integrity, and discipline. Second, with this principle, a person can lead up (the leader's leader), down (the leader's followers), and sideways (the leader's peers). All this can be done through the principle of setting the example, and it costs the leader nothing more than their own will, integrity, and discipline. Ironically, though, the primary effectiveness of setting the example comes from others' behavior. In general, good people will endeavor to emulate those whom they admire or follow those whom they respect. Either way, the leader and the organization end up with people carrying out the desired behavior.

Be technically and tactically proficient.

This principle does many positive things for the leader. First, it allows the leader to contribute to their organization's success, because having a highly developed grasp of what things are necessary to accomplish the organization's mission is what results in success. People who are successful in accomplishing an organization's mission are naturally looked to as leaders—especially if they achieve success on a consistent basis. Second, and maybe more important, is that being in a position of successful leadership fulfills an animalistic need in followers. Followers need to believe that their best leader (one who rises to positions of leadership) will lead them to success over the other guy's best leader. To couch it in caveman's terms: If success equals being able to physically dominate all the beasts and other cavemen that exist, then I as a follower would naturally be comforted by the fact that my group's leader is the biggest, fastest, and strongest caveman there is. And if I feel secure with my leader, I will follow him for two primary reasons: 1) I don't want him clubbing me over the head, and 2) If our group always wins, then we always get to eat, survive, and multiply. In the modern-day world, being technically and tactically proficient in the tasks required to accomplish the organization's mission is no different than being the biggest, fastest, and strongest caveman. The point is that the technical and tactical proficiency of a leader instills in other people a subconscious belief in that leader's abilities and creates a willingness in others to follow that person. The only cost to exercising this principle is the time and effort a leader must invest in learning and perfecting their craft in order to be technically and tactically proficient.

Know yourself and seek self-improvement.

There are two things that a person should know about themself: their strengths and their weaknesses. Those followers who entrust their time, energy, future, and in some cases their very lives to a person, deserve the best leader they can get. This places a great responsibility upon a leader to seek self-improvement at every turn. Before one can even begin to seek self-improvement, they must know themself. While it's almost universally true that we can always get better, even at things that we consider to be a strength, a person only has a finite amount of time and energy in a lifetime. As with anything else, it is incumbent upon a leader to manage this time and energy wisely for maximum effectiveness when it comes to self-improvement. The first step in this wise management is having a good grasp of where you're currently at. This is where knowing yourself comes into play. Once you know where you're at, seeking self-improvement becomes a simple matter of reinforcing your strengths and decreasing your weaknesses.

Make sound and timely decisions.

Sound and timely decisions are something that a leader owes to both their followers and themself. A follower's plans and actions flow from the leader's direction as determined by his or her decisions. A leader owes it to the follower to make the best decision possible with the time, information, knowledge, and insight available to them. The leader also owes the follower as much time as possible to make their own plans and preparations to execute the leader's decision. Planning and preparation contribute to success. More time usually contributes to better planning and preparation. Therefore a leader owes their followers as much time as possible because they owe their followers every opportunity to succeed in executing the leader's decisions. A leader owes themself sound and timely decisions because sound and timely decisions instill confidence in one's leadership on the part of followers. Sound and timely decisions, whether true or not, give the perception that a person believes in what they have decided and knows what they're doing. Ill-founded decisions, waffling, or indecision produces the opposite effect.

Keep your people informed.

For a number of reasons, this is a critical principle for leading human beings and one that is often either overlooked or purposely avoided. First, a follower can't properly execute a leader's intent without sufficient information. Additionally, as things evolve and change during execution (and almost all situations are dynamic ones, so things will change), if the leader happens to be unable to provide further direction, a follower most certainly won't be able to accomplish a leader's intent if they haven't been properly informed of it. Second, when followers find that they are never provided information by the leader but are subject to the consequences of that information (especially if they are negative), they are likely to believe that the leader doesn't consider them important enough or trustworthy enough to know. Any leader whose followers lose trust or faith in them are soon to lose their followers. Third, and maybe most important, is that voids in information, like all voids, will be filled with something. Since human beings have a natural tendency to assign positive attributes internally and negative attributes externally, most voids in information are filled by followers with negative thoughts and assumptions about the intentions of those withholding the information. Leading people is hard enough without letting them come to negative conclusions about the leader's intentions and overall leadership simply because the leader didn't provide them with factual information.

If this principle is so critical, then why is it overlooked or purposely avoided? The answer often boils down to human fallibility. Knowledge, in many cases, is indeed power, but a leader who keeps it from their followers because they believe it keeps them in a dominant position is merely a self-involved, misguided individual. Another reason why a leader may hold onto information is because they believe their followers are lesser in some manner—little people who aren't worth the time or effort to keep informed. Finally, some leaders believe that keeping their followers informed somehow lessens their standing as the leader, giving the impression they aren't in charge. This is a foolhardy notion since it takes more confidence and character to be secure in one's competence as a leader.

Know your people, and look out for their welfare.

Theodore Roosevelt said, "People don't care how much you know until they know how much you care."[2] So why is this principle important? Well, for two reasons. First, at its base level, leadership is a contract between the leader and the led. The terms of that contract are twofold: you do what I say, and I will take care of you. As a cause of nature, people don't naturally find themselves able to care about someone who doesn't care about them. And some level of caring for the leader on the follower's part is always required for there to be a leader/follower relationship. Whether that caring is based upon respect, admiration, or fear, the point is that the follower needs to care.

Second, leadership of each individual is as unique as each individual is unique. Therefore if a leader does not know their people, there is no way they can employ the most effective leadership technique when trying to lead that individual. Also, if a leader doesn't know their people, they most certainly can't look out for their welfare. Welfare is an individualistic thing. If a leader doesn't know their people individually and collectively, they won't know what constitutes the welfare of those people. Knowing someone and subsequently looking out for their welfare takes effort and action; effort and action equate to caring. And as stated previously, there must at least be caring on the leader's part before there is any possibility for caring on the follower's part.

Instill a sense of responsibility in your subordinates.

Following the concept that it is impossible for a leader to know or pay attention to every task required to accomplish a mission, especially in large or complex organizations, instilling a sense of responsibility in subordinates serves three distinct and crucial purposes. First, it acts as a force multiplier. If a sense of responsibility leads to people doing their utmost to do the right thing for the organization and its people, every subordinate that is instilled with a sense of responsibility becomes an extension of the leader, ensuring that the leader's vision, direction, and intent reaches every corner of the organization.

Second, a sense of responsibility in subordinates significantly contributes to an organization being able to achieve maximum efficiency and effectiveness. It is assumed that a leader wants their organization to be the best it can be. If efficiency and effectiveness are measures of success, the more people in their organization a leader has working to achieve optimal efficiency and effectiveness in the things under their purview, the closer the organization will come to reaching its maximum efficiency and effectiveness.

Third, a sense of responsibility usually provides a sense of purpose or is derived from one. Either way, having a sense of purpose is beneficial for the subordinate as a person and the organization as a whole. A sense of purpose directly ties to feelings of contentment and satisfaction in people. Contentment and satisfaction contribute to happiness. A sense of purpose in the individual is beneficial to the organization because it leads to proactive thinking and actions on the part of the individual. The individual will do the right thing because they want to rather than have to or are told to. People doing things because they "want to" adds extra energy and effort to the leader's vision, direction, and intent. Additionally, people who are happy in their work because they are content and satisfied are likely to be more productive and effective, which also benefits the organization and the leader.

Finally, instilling a sense of responsibility in subordinates acts as a system of checks and balances against failure for the leader and the organization. Instilling a sense of responsibility in people piques their interest in the success of the organization. People with an interest in the success of an organization will do their best to ensure that people or events don't jeopardize that success—including any jeopardy to success posed by the leader.

Ensure tasks are understood, supervised, and accomplished.

This principle is part of the glue that allows for independent thinking and action by personnel throughout an organization. The commander's intent governs what needs to get done, while this principle governs the actual doing and the success thereof. Tasks that are not understood cannot be accomplished and are, in fact, not tasks at all. Communication is a two-part exercise that consists of transmission and reception of a message. The last part of this (reception) is not satisfied simply by a message impinging on the senses of the other party. Comprehension and understanding on the part of the receiver is required for true communication to have occurred. Because communication is required in order to coordinate any human activity, communication is an indispensable part of leadership. As a result, a leader ensuring that a task is understood is indispensable. (If you don't believe me, crack open a Bible and read up on the Tower of Babel in chapter 11 of Genesis.)

The supervision part of this principle is just as critical as the others. Supervision assures that the accomplishment stage of the principle happens. Supervision begins with the commander's intent. The commander provides supervision through expression of his intent and desired end state. But as previously stated, most tasks in the realm of human operations require the accomplishment of multiple smaller tasks, and one leader does not have the ability to supervise all of these. This is where supervision occurring at all levels—through having instilled a sense of responsibility in subordinate leaders—comes into play. Having supervision throughout all parts of an organization allows a single leader to control a multitude of people, at different levels, and throughout far-reaching places in an organization.

Train your people as a team.

Few tasks can be thoroughly accomplished by a single person; therefore most require teamwork in some form or fashion. If you expect your people to perform as a team during real-world execution, they first must be trained as a team through practice. Human beings are not "break glass in case of war" machines. While they can seem machine-like as a result of carrying out previous programming, the programming does not occur in a single-entry-type manner like computers. Programming for human beings happens through experiences and teachings. Therefore, to get a reasonably reliable output of actions from a person, they must be programmed with experiences and teachings that are consistent, repetitive, and in line with the desired output. You can't expect a reliable output in terms of performance if you teach one thing but then have the same people experience a different or contradictory thing from what was taught. You can't train them to be one way in practice, then expect them to act completely different when it's game time. You can't train them as individuals and expect them to perform as a team, no more than you can train them like lambs in peacetime and expect them to be lions during combat. So if the goal of an organization is to accomplish a set of tasks, and most tasks require teamwork by a group of people, it logically follows that you must train your people as a team.

Employ your unit in accordance with its capabilities.

Not following this principle is both foolhardy for the organization and foolhardy for its people. It's foolhardy for the organization in that not employing a unit in accordance with its capabilities risks failure of the mission to be accomplished. And purposely risking mission failure is both reckless and displays a lack of intelligence. Violating this principle is irresponsible for your people because it is tantamount to setting them up for failure. Note that it is critically important to understand the answers to these two questions when employing this principle: First, "What are the *current capabilities* of the organization and the people within it?" It is impossible to employ the organization and its people in accordance with their capabilities if you don't have intimate knowledge of them. Second, "What are the *potential capabilities* of the organization and the people within it?" This is important because it is a leader's responsibility to grow and develop their people. Growth only happens when a thing is stretched outside of its current capabilities or comfort zone.

Seek responsibility and take responsibility.

Seeking responsibility is demonstrative of a leader. Taking responsibility is demonstrative of a person of good character. Why are these two attributes important? First, leadership is more felt and sensed by people rather than being rationally explained. When a leader seeks responsibility and not just authority or power, then their followers are more likely to develop confidence in (as well as respect for) the leader. Successfully handling responsibility sets the example amongst their peers and instills a sense of trust in that leader amongst their followers.

CHAPTER TWO: Exemplary Traits

The things that make a good leader must be a part of their makeup.

Leadership Traits

Justice

Positive justice is where the leader ensures that people get anything that they deserve that is beneficial to them. Negative justice is where the leader ensures that people get any negative consequences that they deserve for their actions. Leaders need to ensure that justice is upheld, both positive and negative, in as many situations as possible. They also need to ensure that the lower-level leaders are being just in their actions as well. The first of these needs to be done for the organization. The second supports the followers' perception of the individual as a leader. Teamwork is a social construct and entails a social contract (most times informal and unacknowledged) or sorts. True teamwork requires individuals to willingly submit themselves to the team. For this submission to genuinely occur, the individual must believe they will be treated equally and fairly. Justice represents the equal and fair treatment of all individuals. Therefore the leader must ensure that justice occurs throughout the entirety of the organization and its operations. Once people lose their belief of fair and equal treatment, they lose their commitment to the team, the organization, and to the leader. The other part of the justice equation is that the leader must also act justly. If followers do not see the leader consistently act in a just manner, there can be no faith on their part that they will be treated fairly and equally by the leader or organization. And without that faith, there is no commitment.

Judgement

Judgement is what separates knowledge from wisdom. Judgement is not simply knowing stuff; it's being able to call upon and apply one's intelligence, experience, morals, and values to a given situation in order to arrive at the wisest decision possible. A leader can only hope to reach the level of wisdom of King Solomon (if you don't know who that is, crack open a Bible). However, while few will ever reach that level of wisdom, the pursuit of it is a worthy and necessary endeavor. Much of a follower's willingness to comply stems from Maslow's hierarchy of needs, which explains the stages of behavior motivation. A concise summation of the hierarchy says, "First, I will do (in this case follow someone) what will keep me alive and healthy. Second, I will do what will make me happy."[1]

Because of this mindset, followers may not like a leader's personality but will choose to follow them if the leader is successful in demonstrating their ability to contribute to the follower's survival and happiness. A leader's judgement is a critical indicator of their leadership abilities. Judgment is a tool. If a follower trusts their leader's competency in using that tool, then they will willingly follow that person—even when they don't agree with a leader's decision, or when they don't understand the leader's decision, or when there isn't time for the leader to communicate what the decision is. This last one is particularly important in life-or-death situations. Demonstrated good judgement puts credit in a leader's bank account that they can draw upon with followers in emergency situations. The more strenuous, hazardous, and critical the situation, the larger a withdrawal a leader must make from their account. Without having made steady and sizeable contributions to the account, a leader could find themself with insufficient funds at the time when they most need it. In the Mann Gulch fire in 1949, a smokejumper crew's foreman demonstrated how critical having good judgement is, as well as demonstrating good judgement to one's followers.[2] In this case, the foreman failed to demonstrate it to his followers before they became involved in an emergency situation, as a result, his followers did not follow his lead at a critical time when he needed them to trust that he knew the right thing to do. His leadership failure contributed to many of them perishing.

Dependability

I once attended a lecture by someone whom I consider to be one of the great thinkers of our time, Simon Sinek.

During the lecture, he talked about how human beings are still very much governed by their animal nature and the social actions we take as a result of this instinctive nature. He defines one of these social actions as grouping together for protection. One thing that occurs with this assembly is that the group identifies, defers to, and supports a leader in their position of privilege. In exchange for this deference and support, the group expects that the leader will provide for and protect the group. This is the social contract between the leader and the ones being led. Part of that contract, just like any other contract, is the assumption of good faith on the part of both parties who must believe that each side will fulfill their contractual commitments. If any party to the contract is not confident that the other party will fulfill their obligation, then there is no meeting of the minds and therefore no contract exists. This is where dependability comes in. A leader has contracts with three distinct groups: his leader, his peers, and his followers. All three of these groups must be confident that that individual leader can be depended upon to fulfill their part of the contract. If a person ever loses the confidence of those groups and cannot be seen as dependable, then their contract is void, and that person has no standing as a leader within those groups. Once it is known that a person cannot be depended upon, potential followers will never put him or her in a position or situation where their participation would be critical to the success of anything.

Initiative

Initiative is a foundational part of the makeup of a leader because being a leader requires initiative in and of itself. You can be in charge of someone and not be their leader. Leadership requires being proactive. Initiative is demonstrated through proactive action. Leadership is hard and requires a person to go against normal human tendency. In his book *Challenging the Law Enforcement Organization*, Jack E. Enter says that most leaders fail because leadership is contrary to human nature, and you won't see it modeled or encouraged by peers.[3] Initiative in people is one of those things that closely follows the laws of physics. Physics says that an object in motion will stay in motion, and an object at rest will stay at rest unless acted upon by an outside force. Because leadership is not easy or natural, the most natural state (rest) is to not engage in it. The outside force needed to change this state is a person's will and determination. That force, in many instances, takes the form of initiative, which requires action. Initiative is demonstrable of "giving a damn," and giving a damn can't be faked. This is important because if people know that you give a damn about the organization, then the people—and the mission—are more apt to succeed because they followed you, the leader who actually cared enough to do something.

Decisiveness

In the movie *Wyatt Earp*, Earp directs one of his deputies to go confiscate the weapons of some drunk cowboys who are in town because people weren't allowed to carry guns while in Dodge City. The deputy stops the cowboys and tells them to hand over their weapons. The cowboys refuse to obey the deputy, and the deputy continues exchanging words back and forth with the cowboys as the situation quickly heads toward a violent confrontation. Earp quickly enters the scene from behind the cowboys and pistol whips them into unconsciousness, then takes away their weapons. Oh, by the way, one of the cowboys was setting up to shoot the deputy with a secreted gun just before Earp intervened. Later, Earp tells the deputy he needs to find another line of work. Earp tells him that (because of his temperament) he could get himself and others around him killed; that the land is harsh and it doesn't suffer fools. When the deputy protests Earp's assessment, Earp tells him straight out that his problem is that he isn't a deliberate man; that he is too affable.[4]

One of the worst things a person trying to get others to follow them can be is someone who waffles back and forth on matters, or a person who flitters about in their thinking and deliberations, or a person who refuses to act once a decision has been made. None of these things inspire confidence in other people that they should follow you. Just like the cowboys in the movie, people won't listen to you. Even worse, some will develop contempt for you based upon you not being able to make sound and timely decisions. Decisiveness doesn't mean making and executing rash or foolhardy decisions. What it does mean is that in any situation there is a finite amount of time to address the situation. That time must be strategically divided between deliberation, planning, and action. Each situation is different, so the amount of time that can be dedicated to each of these phases varies. A person should use as much of that finite time as possible for the deliberation and planning phases, as these make the execution phase infinitely easier. But once a person reaches the end of their deliberation time, whether by their own accord or by forced circumstances, they must move forward to the planning phase. And once they have reached the end of whatever time is available for planning, they must move forward to acting. The steady advancement through all these phases reveals if a person is decisive or not.

Tact

Tact is an indispensable tool for the person trying to communicate, especially when they are transmitting a message that will likely be unpleasant to the receiver. In the transmitting-receiving process (communication), it is the responsibility of the sender to ensure that the message is received and understood. Ensuring the receiver has the intended understanding of what was transmitted is as critical to proper communication as all the other parts, but it's most often where communication failure occurs. This is because understanding is largely dependent upon individual human interpretation. The first parts of communication merely require a message to communicate, a form to communicate it in, a means to communicate it, and a means to receive it. The last part requires a focused effort on both the sender's and receiver's part. The sender, to construct and transmit the message in such a manner that it has the best chance of being interpreted exactly as the he or she intends. The receiver, to diligently process the message so that it has the best chance of being correctly understood. Constructing the message in a manner that the receiver will willingly show diligence in processing it and understanding its meaning is where tact comes in. Having the best message in the world means nothing if you alienate the receiver through your communication style. Tact is a tool that is applied to a message to give it the best chance of it positively impacting the sender. Tact is most often needed when a sender is trying to communicate a message that is likely to be considered negative by receiver. Negative messages easily get a receiver's defenses up, which automatically hinders communication. The purpose of tact is not to change the underlying nature of a message, but to make it more palatable to the receiver and thereby more likely to be taken under consideration and acted upon.

Integrity

Integrity is what you do when no one is looking and when everyone is looking! You have probably heard the first part of this quote before but not necessarily the last part. Both are applicable and equally important. To differentiate between the two, the "when no one is looking" type of integrity originates from a person's character. There is no impetus to exercise it other than one's own morals, values, and principles. Additionally, there arises the difficulty of going against one's own nature because exercising it usually entails a self-inflicted hardship. The "when everyone is looking" type of integrity is also internal in origin but often with an external impetus in exercising it. For instance, when a subordinate makes a mistake (especially a public one), and discipline or reprimand is in order, it is human nature for a superior to want to satisfy the desire to demonstrate their power and authority by carrying out the reprimand or discipline in a conspicuous and public manner. However, it takes the exercise of integrity to do what is right in such cases, which is dispensing with the matter in a way that is best for the organization and the subordinate. People in positions of authority and power must work harder and pay closer attention to having integrity than people with less influence. This is because authority and power have two pitfalls that a leader can easily step into. The first is that power and authority make it easier for a person to sidestep, ignore, or manipulate rules, regulations, customs, morals, and norms that others are required to follow. The second is that there are usually fewer people willing to act as an external system of checks and balances for the leader, and fewer mechanisms (audits, reviews, cosigns, etc.) that exist to serve that purpose. Because of the inverse relationship of the existence of external checks and balances versus level of power and authority, integrity is a constant pursuit, like a muscle that must be constantly exercised, less it atrophies and dies.

Endurance

"The race is not [given] to the swift . . . but to the one who endures to the end" is found in Ecclesiastes 9:11 and Matthew 10:22. It's a truth I have clung to over the years when I need a boost. I first implemented this directive in childhood, after learning an important lesson in high school gym. By far, I was not the fastest runner and still am not. However, I noticed in gym when we had to run laps as part of our exercise, the naturally fast guys took off and completed the run long before I could. I began slow and finished long after the fast runners, but I always finished! The majority of everyone else started out as fast as they could, thinking that meant they would finish faster. But what actually happened is that as the run the progressed, they slowed down and, one by one, I passed many of them. It was at that time I reasoned that while I may not finish first, I certainly wouldn't be last, and I would finish ahead of some people who initially looked like they had left me in the dust. The only thing required of me was to relax and run my race.

Someone once asked me which leadership trait was most significant to me. After thinking about it, I couldn't find one of the fourteen that I held in higher esteem than *endurance*. While I can't classify it or any of them as the most important, I've found that it represents what I had done for a large part of my life and the experiences therein. Ironically, it is a trait I have rarely heard cited as being at the top of anyone else's list. Over the years I have encountered people who were faster, stronger, smarter, and better skilled than me. People whose level of excellence and ease impressed me. I was privileged and am honored to have served with or have been associated with them. But I succeeded beyond many of them in the same endeavors largely because I endured. Endurance is a key trait for not only the leader to have, but also for the effective leader to develop in their subordinates.

A leader can never be hot or cold, hungry or thirsty, tired or unsure; this is where endurance comes in. Because leaders are human like everyone else, it is inevitable that they will be all these things at some point in time. Endurance is the character trait that allows a leader to override, or at least cover up, those fragilities of being human that, if shown, could lead to their followers acquiescing to their own fragilities, resulting in failure to accomplish their mission. Therefore it is also important for a leader to make a concerted effort to develop endurance in their subordinates.

Endurance is a muscle that must be developed and maintained like every other muscle. It requires active resistance to the common detractors from human performance. Active resistance is hard work, and it is often much easier to just give in. With most things, the higher the stakes, the more detractors that will present themselves—and the more severe they will be. If one does not build up their endurance in order to be able to carry on in the worst of times, their will to persevere will fail them when it really counts. Endurance is not just a physical thing; it's mental as well. Endurance is a mindset. It is a state of belief that no matter what the hardship is, I can and will deal with it. No matter how hard, long, complicated, or unpleasant the task, we as a team will get through it successfully. This mindset is not something that can be imparted or developed overnight. It must be purposefully cultivated, reinforced, and maintained over time to be truly ingrained within a person's character.

Bearing
Bearing is a highly beneficial trait that is a total freebie. I say it's a freebie because it requires nothing more than a leader's conscious effort to project and maintain it. Human beings are creatures of perception. What we perceive is our reality. The good and bad thing about that is what we perceive may not necessarily be the truth. The

bad part is we can be sure about something and be totally wrong about it. The good part is that you don't necessarily have to feel or believe the thing you are presenting by having bearing, because it has the same effect on those around you whether it is real or not. In other words, because it is largely based on others' perceptions of you, you can achieve the desired effect on those around you even if you fake what is perceived as bearing. Bearing is made up of two things. One of those things is the ability of a person to consistently project qualities and traits that would cause others to look upon them in a favorable light. The other thing is the ability to compose and comport oneself in times of stress and strife, the same as you would in times of calm and ease. Martin Luther King Jr. said, "The ultimate measure of a man is not where he stands in moments of comfort and convenience, but where he stands at times of challenge and controversy."[5] Bearing is how you appear to others when you're standing in those times of challenge and controversy.

Why is bearing important? In the movie *U-571*, Matthew McConaughey plays a young U.S. Navy officer who has been unintentionally thrust into commanding a German submarine. The submarine was captured as part of the important mission of getting ahold of an enigma machine so that the allies could break the German's encryption code during World War II. Unfortunately, McConaughey's own ship is sunk during the process of capturing the German boat, and instead of taking the enigma machine back with them, McConaughey and the few sailors who accompanied him on the mission are forced to try to and return home, utilizing the German submarine. Actor Harvey Keitel plays the senior enlisted man on board the submarine during the mission. During one critical point in their attempt to return home, Matthew McConaughey has a situation where he doesn't know what to do, and all the crew's lives are on the line. He loses his bearing while trying to figure out what to do in consultation with the crew. McConaughey then makes the mistake of verbally stating to the crew that he doesn't know what to do. Harvey Keitel takes charge of the situation and straightens the crew out by force of will and outright threatening to beat the hell out of them if they don't follow orders. Later, Keitel sits down with McConaughey and tells him in plain terms that he is the captain, which means he can never *not* know what to do.[6] He must always know what to do because if he doesn't, fear will break out among the crew—and fear kills.

Having and maintaining one's bearing fulfils both the animalistic and intelligent nature for our human followers. In the case of the animalistic side, we look to others for social cues about our own well-being and how to act. If the leader appears confident and unconcerned, the rest of the pack naturally assumes that it is safe as a result of being protected by the leader. In the case of the intelligent side, we look for social cues to make logical connections and decisions as well. Once again, if a leader appears calm, confident, and in control of themselves, followers will logically believe that he or she is in control of the situation, sure of their ability to handle everything and confident in a positive outcome. This makes it less likely that followers will lose control of their emotions and faculties, which, if allowed to happen, would significantly increase the chance of failure. In the case of peers and superiors, maintaining your bearing during the worst of times, as well as the best of times, goes a long way toward gaining the respect and confidence of these two groups of people. For superiors, peers, and followers alike, a leader's display of bearing is a sure deposit in the bank of confidence, which yields untold dividends in other areas.

Unselfishness

Leadership requires unselfishness. In Freud's psychic model of the animalistic id, the moralistic superego and the ego that governs between the two, selfishness, which we possess from the beginning of our existence, falls squarely into the id's ballpark. Unselfishness, which is developed through social programming, resides in the realm of the superego. The ego is what we develop as a mature adult with our own concept of the world.[7] Leadership is a conscious, intelligent endeavor and is therefore ruled by the ego part of the psyche. Leadership itself requires unselfishness in two different ways. First, it requires a person to be unselfish enough to be willing to be a leader. As was stated earlier, leadership is not a natural attribute to humans and it is not an easy undertaking by any stretch of the imagination. The second way leadership requires unselfishness is that a leader must be willing to do unselfish things in order to practice leadership. A leader must be willing to suffer and endure hardships. As many times as not, a leader's hardships are self-imposed, or at least aren't actively avoided. Unselfishness tells followers, peers, and superiors a few things about a person. Unselfishness says that you are just, possess good morals and values, and recognize other people's value outside of what benefits you. It also shows people you think yourself no better than them as a human being where it concerns accomplishing the mission. Just as importantly, the demonstration of unselfishness will cause others to be unselfish toward you and inspire them to demonstrate unselfishness toward all others. Jesus Christ, the ultimate exemplar of leadership, demonstrated the paramount example of unselfishness by willingly assuming our sins and allowing himself to be sacrificed in order to rectify all of mankind's relationship with God.

Courage

In my career as a SEA, I have questioned officers on why they would not give a Marine a marking in the courage block of their fitness report. Most officers would mark that block as "Not Applicable" unless they were doing a combat fitness report. I thought that was totally uninsightful of the nature of the Marine Corps as a human organization. When you think about it, combat courage (when bullets are flying and bodily injury or death is a real possibility), or the lack of it, is relatively easy to recognize and judge. Actually, actions taken during an immediate life-or-death situation is the simplest type of courage to discern. However, it is possible to observe whether somebody has courage in other ways and areas. The more complex type of courage, and the one that is overwhelmingly more often needed, is social courage. Social courage comes in different forms, but the basis of it revolves around human interaction. This courage is exemplified in instances such as a person being willing to speak truth to power even when the truth is unpleasant; a person being willing to tell their peers and superiors that they are wrong even at the risk of alienating them; or a superior having the guts to admonish or discipline a subordinate in a direct and face-to-face manner. People use or fail to use this type of courage every day, and sometimes multiple times a day. Because the requirement to exercise it is much more frequent, and social conditioning tends to dissuade us from it, social courage is equally as arduous as the getting-shot-at type. It is obvious why the combat-type of courage is necessary of a leader. More battles, campaigns, and wars have been won by the courage of an individual or a few people at critical moments than all the grand plans drawn up by all the great generals in the world. What is not quite so obvious is the need for social courage. Social courage helps ward off faulty thinking, nix bad ideals, and collectively hold leaders, followers, and peers accountable.

Knowledge

My father used to say all the time that, "Knowing always beats thinking." Knowledge is the key component in fulfilling the leadership principle of being technically and tactically proficient. Knowledge is not simply the attainment of information. Rather, it is the attainment of information, the comprehension of that information, and the understanding of the applicability of that information to a given task or situation. The great thing about knowledge for leaders is that it does not have to be firsthand. Knowledge can be gained through academic study and training. Therefore one must be a scholar of their profession in order to call themselves a professional. The acquiring of knowledge, the maintaining of it, and the pursuit of it is even more is critical to current and future mission accomplishment. Knowledge contributes to the success of the mission not only through the direct individual actions of a leader, but also in the direction he or she provides to others in accomplishing tasks associated with the mission. Knowledge also contributes to the mission through the confidence that superiors, peers, and followers have in a leader's ability. Superiors who have this confidence will allow the leader to lead; peers who have this confidence will emulate the leader;

and subordinates who have this confidence will follow the leader unquestionably.

Loyalty

Loyalty is a double-edged and multifaceted sword. It's double-edged in that it can only be garnered through the giving and demonstration of it by the person who wants to receive it. In this way it is very similar to trust and actually derives from trust. Loyalty is given when the person trusts that another person or thing is deserving of it—or at least trusts that giving it will benefit them. Similarly, if a leader is looking to garner the loyalty of their subordinates, he or she must have their trust in one of these ways. There are many ways to garner the trust necessary for loyalty; however, we will go in depth on trust later. Loyalty is multifaceted in that there can be loyalty to a person or to an entity, such as an organization, and at times there are competing interests when loyalty to both is involved. Also, loyalty to the person or the organization may change, depending upon the time or the situation. Loyalty to a person, or lack thereof, usually centers around moral or ethical issues. Whereas loyalty to an organization usually centers on rules, regulations, policies, and professional ethics. The spirit of the law and the letter of the law can sometimes clash with one another where it involves loyalty. And sometimes what's right at a specific time or situation can clash with both of those other two things. Regardless, loyalty is a primary ingredient in the bond that allows society and civilization to exist. While it may not affect the coming together of people, it most certainly affects whether they stay together. As stated earlier, most tasks require more than one person to accomplish. Therefore tasks require people to come together and stay together, at least long enough to accomplish them. The end purpose of human beings' categorization of themselves into family, cultures, country, and race is a determination of where loyalty lies. It is a major factor in group norms and is, as often as not, a controlling factor of human behavior more than other social issues. Loyalty is critical to any group or organization's success and, as a result, is critical to any leader's success. This is well-known in the commercial and marketing arenas and has even gained a modern-day name for itself with the term "branding." Organizations go to great lengths to brand their product. Why? Because brand loyalty equals continued company existence and success.

Enthusiasm

On one of many cold mornings in my military career, I was up with my company for a morning PT run. The company was formed up outside, and I was going out to join them. While approaching the formation, I passed one of my staff sergeants—who had his arms crossed, his body balled up as tightly as it could be (and still be considered standing), and was visibly shivering. His miserable appearance and sullen demeanor at that moment in no way exuded the high standards I expected of a staff non-commissioned officer, and certainly not that of a leader of other people. I stopped and asked him if he was cold. His response was that he was freezing, which displeased me even more. So I explained to him (in a not-so-nice first sergeant way) that no, he was not cold. I explained to him that if he was cold and miserable, that meant the troops would believe it was okay to be cold and miserable. And if they were cold and miserable, that meant they couldn't perform in an optimal manner. And if they couldn't perform optimally, then they couldn't accomplish the mission. And if they couldn't accomplish the mission, we'd lose, removing all purpose for any of us. For added perspective to this story, I must tell you that it *was* freaking cold that morning. We were only wearing running shorts and a T-shirt, and I myself was freezing my ass off. The only difference between the staff sergeant and me was that I made a concerted effort to not *show* that I was cold. Enthusiasm, for me, is one of the most important leadership traits there is, yet it is one of the most overlooked and least-cited ones.

You know the old saying that no one likes a sourpuss? Well, I will tell you that not only does no one like a sourpuss, but a sourpuss cannot lead people. Enthusiasm in large part stems from optimism. Whether you are optimistic that things will get better or that you can make things better matters not; just the fact that you have a hopeful belief that motivates you in an outwardly positive direction is enough. I say outwardly positive because you may not necessarily truly feel enthusiastic. But as one of my favorite sayings goes, "Fake it until they feel it." Why is this? Why am I encouraging a leader to fake anything? Because enthusiasm is one of those things where unselfishness must be exhibited. People say that enthusiasm is contagious, and it is absolutely true. Not only is enthusiasm contagious, it is beneficial to people's overall well-being, and as such, it is beneficial to mission accomplishment. Therefore a leader has a duty to provide enthusiasm for his or her people whether he or she feels it themselves or not. Additionally, one of those little quirks of human psychology is that if you fake enthusiasm long enough, you do eventually feel it. But regardless of whether you end up feeling enthusiasm or not, it is a key ingredient of a team's success and mission accomplishment and therefore a leader is duty-bound to outwardly project it. Enthusiasm requires nothing and costs nothing to exercise. Remember: whether the result of an endeavor is good or bad, no one can ever chastise you for having approached a task with enthusiasm and having enthusiastically tried to accomplish it. Conversely, pessimism is also contagious. People who do not fully believe in a goal or in themselves will most likely not accomplish anything. As Henry Ford said, "Whether you believe you can do a thing or not, you are right."[8] With true enthusiasm, a leader can will his or her people to achieve things that even they might not believe possible. As such, enthusiasm is a powerful tool for a leader to have in their toolbox. And because it is free, any leader would be a fool not to use it.

CHAPTER THREE: Contributors to Leadership

Good leadership can overcome a lack of favorable conditions; the reverse is not true.

While leadership can be an abstract thing and hard to define in some respects, there are certainly aspects that contribute to being successful at it. These contributions to successful leadership can be defined by things a leader should do and things a leader shouldn't do.

Commander's Intent

Give the commander's intent and allow for different approaches to execution; otherwise, do it yourself. People aren't robots to be minutely programmed, nor are they avatars to be used as remote-controlled substitutes. If you attempt to utilize them that way, you won't be successful in achieving your mission, nor will you get their best efforts. Tell them what you want done, but minimize telling them how to do it. In giving the commander's intent, a leader should go back to elementary school basics by clearly articulating who, what, when, where, and why in as concise a statement as possible. However, the *how* should be given in generalization not specificity, if at all. The exception to this is if the leader genuinely believes that the situation and success of the task requires it to be done a certain way and no other. Commander's intent allows people to accomplish tasks in the absence of constant guidance from the leader. Subordinates knowing and understanding the commander's intent ensures that they will be able to make sound and timely decisions when necessary, then take decisive action where appropriate. Having the commander's intent also gives subordinates the freedom to generate ideas and courses of action that they have ownership of. This approach almost always leads to better execution and is more likely to produce better results than what the leader had in mind.

Too Many Priorities = No Priorities

Develop a few critical objectives and focus on those. A wise man once said, "Too many priorities are no priorities at all!" Everything being a priority is like the boy who cried wolf. After a certain number of false reports, people will stop giving them the attention that is due a true priority, and eventually that will extend to the person who keeps coming up with the priorities. One unintentional way leaders fall into the "everything is a priority" trap is unnecessarily delving into the minutia of things. If you are a leader or in a position of authority, your attention on something automatically raises its priority among people below you. If you as the leader pay the same attention to everything, then you make everything a priority. Another way leaders fall into the trap is by believing anything and everything can be done regardless of the limits of their resources; or that everything can and should be focused on with the same level of intensity. Both of these things aren't true, of course. How do we know this? Simple. There are limits to human existence, which means there are limits to everything humans do during their existence. A leader's job, for the most part, is to draw the "big blue arrows" of vision, direction, and intent (VDI) for their followers and then provide supervision, guidance, correction, and clarification (SGCC) as necessary for their followers to execute. A leader's priorities will then be naturally identified by the things that are crucial to supporting their VDI, and the leader will make those priorities clear by focusing their SGCC on those things. Not that a leader should forget about or ignore everything else that concerns the operation of their organization. It's just that when it boils down to it, a leader (such as a sports coach) is just another position player on a team. And just like every other player on the team, a leader must trust his teammates to play their positions well enough to support winning.

Find Good People. Help Them Find Their Right Place. Let Them Do Great Things.

Identify good people, put them in the right place to maximize their utility, then give them the authority and leeway to do good things. This concept contains many other sub-concepts that are very important to leadership, such as, "You don't have to be the smartest person in the room; you just have to be smart enough to get the smartest people into the room." The fact is that the art and science of leadership comes into play in this area as much as any other. Identifying good people can be a science in that you can evaluate people on the objective points of training, experience, knowledge, etc. On the other hand, it can be an art as far as sometimes people are in the wrong position for what best suits them, or sometimes people don't know their own strengths, and you must ascertain them. Putting people in the right place can be scientific in that you merely look for the area that can most benefit from their knowledge, experience, and training, then put them there. However, it can also be an art in that sometimes the best place for people is not the place where it would seem their attributes would best fit. Sometimes the best place for them is determined by other intangible things that must be deduced. Turning them loose with the authority and leeway to do great things is multifaceted. The authority part is scientific in that you make known in a formal manner that a person has say-so over certain things. However, the leeway part can be very much an art in that you must provide supervision while giving people freedom of action to execute. The art part is where you try to determine what ratio of these two things to give in different situations and at different times.

Finding the right people for the job not only makes sense for the organization; it makes sense for the people as well. The organization gets optimum work efficiency, and the person gets the happiness from doing something that they're well-suited for. Putting the right person for the job in the right job is not only important for the organization to get the maximum benefit out of its most expensive resource, it also is beneficial to the person in that they will likely be most effective in a job that is well-suited for them and will likely have the easiest time of being effective. Giving the right person in the right job the authority and leeway to execute that job makes sense for the organization once again because you want to maximize the utility of the resource, in this case the person. It makes sense for the person in that not having the authority to do a job that they're responsible for accomplishing is setting a person up for failure. Not having the leeway to do their job diminishes the persons effectiveness in accomplishing the task. In both aspects, this effects the person negatively in the form of unnecessary frustration.

Never hold anyone responsible for something they weren't provided the resources or authority to accomplish.

To hold someone unjustifiably responsible is the essence of setting people up for failure. A person who behaves in this manner loses their moral authority with their leaders, peers, and subordinates alike. People will know and will look unfavorably upon you when you unjustly punish someone for something they had no control over it. People will look at you as an unfair person, and once they see you in this light, even for one thing, it will carry over into everything, and your ability to lead those people will be severely diminished. This is because people don't like unfair treatment, and they don't like people who treat others unfairly. Even when the people being treated unfairly isn't them, it's hard to watch because they have no reason to believe you wouldn't do it to them.

Don't set people up for failure, but do put them in positions that will cause them to stretch and grow.
Setting people up for failure most often involves tasking someone with doing something when it is not known whether that person has the knowledge, skills, or abilities to do the thing, or asking someone to do something and holding them accountable without providing them the resources, support, and authority to accomplish the task. Setting someone up for failure is much different than setting someone up for growth and development. Setting someone up for growth and development is tasking them with something that will require them to use the maximum level of their knowledge, skills, and abilities. They might even have to obtain new knowledge, skills, and abilities to accomplish the task. The only caveat to that last part is that you must make sure that the new

knowledge, skills, and abilities they will have to obtain are within their grasp and ability to obtain. In other words, if you're expecting them to stretch, you better make sure that they're loose and limber enough to stretch the distance that you're asking them to. The difference between setting people up for failure and setting them up for growth and development is whether you tailor your provision of authority, resources, and support in order to make them a better leader. For example, setting someone up for failure is once again holding them accountable for something you didn't give them the necessary authority in order to ensure success. However, setting them up for growth and development would be giving them constrained authority. Constrained authority is my term for just enough authority to accomplish the task, but not enough to make it easy for them to accomplish the task. For instance, it's easy for a guy who's in charge of all pieces of the mission to control the action and outcome. It is a lot harder and causes a lot more work for that person if you only give them a piece of the mission and require them to work harmoniously with others who have authority over other pieces of the mission in order to successfully accomplish the mission. One can look at resources in the same matter. Not providing somebody the basic resources necessary to accomplish a task you assigned them is setting them up for failure. Providing them with limited resources beyond what is absolutely required to accomplish the mission versus providing them with the optimal amount of resources is the difference between just ensuring mission accomplishment or ensuring growth and development. Support can be thought of in the same manner as the other two. Assigning someone a task and then making it clear by your actions that you don't support them is, in essence, setting them up for failure. On the other hand, letting them know you will support them, but only when matched by them putting forth their maximum effort, ensures they will be successful and sets them up for growth and development.

Put your ego away.

No, you're not the only one who can accomplish the mission or do that thing right! The belief that this is true is the height of arrogance and hubris. It is often demonstrative of a serious individual character flaw. Sometimes the flaw is that a person needs to be viewed by others as the one who holds everything together and gets stuff done. Sometimes the flaw is that the person needs to believe this is true about themselves. Still, other times people use it to justify their bad behavior. Regardless of the reason why, this type of thinking leads to many things that can subtract from providing optimum leadership because those who think this way leave no room for the possibility that they could be wrong. And if they have no capacity to consider that they could be wrong, then their internal and external checks and balances against human nature won't work. Another problem with this type of thinking is that if you believe you're the only one who can do it correctly, you'll never task someone else with doing it. Which means you won't be able to delegate. And since a leader cannot do everything themself, one who cannot delegate will do all things in a suboptimal manner and ultimately fail somewhere. Finally, if you think that you're the only one who can do it correctly, you'll never give others room and opportunity to grow, which is also a failure in the leadership department.

Give your followers room to execute.

Yes, this gives them room to make mistakes and to fail. But it also gives them room to innovate, to excel, to practice and refine through trial and error, and to lead themselves. That last one is particularly important because a leader's ultimate goal is to have every individual being a leader in their own little grid square. And the first leadership grid square that everybody has responsibility for is themselves. The wise leader will of course make sure that the room they give is done in a controlled manner. For instance, the wise leader is not going to give a follower room to screw up in a no-fail situation, or give a follower so much room on a time-sensitive task that the task can't still be accomplished if they fall short. This may seem like double-talk. As if I'm saying on one hand to give a follower room, then saying to remove room because they may fail—but this is not so. The majority of things in life don't reach the level of there being absolutely no room for error in getting the task accomplished. Most often, leaders don't want to provide the room because it takes extra time and extra effort to do so. Or because they believe that anything that doesn't ensure perfection is a negative thing. But some of the best and longest-lasting positive lessons are derived from mistakes and failures. Additionally, without room for failure, there isn't any room for success and excellence. So if you don't give people the opportunity and room to execute, they won't learn, they won't develop, and they won't excel.

Maintain your North Star.

Stayed grounded in your foundational beliefs. Passing fads, new ideals, concepts, and ways of doing things come and go. Some add granularity to what is already known, but the majority add little or nothing to the pool of meaningful existence. In the great movie *Pure Country*, George Strait plays a country music star who gets sick of the business and runs away to the country, ending up in small town and finally at the residence of a good Samaritan who takes him in after a drunken night at the local bar. The morning after his night at the bar, Strait is talking with the good Samaritan's old father at breakfast table. The father tells Strait that in his opinion, people talk too darn much; they are always yapping about something when they should be doing something productive like eating, sleeping, or working. The father goes on to say that what is there to be said that everyone doesn't already know, right? Strait responds to the father's opinion with an "I guess." The father stares at Strait with a confounded look and says what is poignant to this leadership point. The father says that a man should never guess, that he's gotta know what he is doing because guessing leaves a person wide open to suggestion, which is why the country is going to hell . . . everybody's open to suggestion.[1]

I'm frequently reminded of that scene because there is an absolute truthfulness about what the old man says. There are some things that a person has to believe—deep down in their bones, with unassailable conviction, as sure as they know their own name. Things they know, that they know, that they know. Maybe an even more poignant point made on this is found in Matthew 13, when Jesus speaks of the seeds that fall by the wayside being devoured by fowls that come by. The seeds that fall in stony places spring up fast but are shallow and soon scorched by the sun, withering because they have no deep root system. And the seeds that fall among the thorns are subsequently choked by the thorns. The point being is that if you are not rooted deeply in something, you will neither survive nor be fruitful. A leader has to determine to the best of their ability what is right. Once he or she has determined this truth, a leader has to believe to the point of certainty that they know, that they know, that they know that those things are right. Those are what I call foundational beliefs. Once determined, a leader should cling to those beliefs like he clings to his next breath. They should be so ingrained in him that when he doesn't know what to do or has to make a quick decision, he is able the fall back on those beliefs and make a decision according to them, knowing that whatever the results, he is satisfied with the basis upon which he made the decision.

Being able to lay your head down at night and sleep comfortably with whatever you did do or didn't do that day is critically important to a leader's physical, mental, and spiritual well-being. If a leader makes decisions and takes actions based upon their foundational beliefs, then no matter the outcome, it is highly probable that they will be able to sleep peacefully at night. As a matter of fact, a leader's foundational beliefs should be a guiding factor in all of their decisions and actions. If those foundational beliefs are solid and true, then they should be applied to and provide sound reasoning in every situation. However, while holding on relentlessly to their foundational beliefs, a leader should always be open to the discovery, enlightenment and growth that they provide as well. Because while foundational beliefs should almost never change, how they translate and interact with the real world will likely change from time-to-time.

Never lose sight of the organization's primary purpose.

Leaders sometimes get bogged down in the rigmarole, details, and constant problem-solving and lose sight of the reason for it all. Most decisions made from this point of view, if not immediately wrong, are wrong for the long-term future of the organization. To share a real-life example, there was a communications battalion commander who was looking to revamp a new support program that had been instituted across the organization. After having the program revamped, the commander gathered his staff and those in top leadership positions in the battalion's companies to have a meeting and outline how the revamped program was going to be executed. The revised program was briefed, and the entire audience came to the same conclusion. The manner in which the new program was to operate would consume an incredible amount of training and operational time (leadership time and energy); and while its purpose was to support the unit, the proposed manner of execution would detract from the unit's primary mission. After the program's operation was briefed, one of the company commanders in the room stood up and inquired of the battalion commander what things did he expect/want the companies to stop doing in order to execute the program in the manner prescribed. The battalion commander appeared to take offense to this question and asked the company commander what he meant by the question. The company commander followed up with quite rightly stating that there was only a limited amount of time to do everything, and if that if he wanted the program to be executed in this manner, then time had to be taken from other obligations, such as training and operations. For this particular case, the company commander was absolutely correct. The manner in which the revised program was devised to run would require a ridiculous amount of time and energy on the part of the organization and its personnel when considered against the purpose behind the program. Worse yet, it would be self-inflicted pain. Unfortunately, rather than considering and addressing the legitimate question, the battalion commander just became angrier. I would like to believe it was anger and not just foolishness, but the battalion commander's response to the company commander and everybody in the room was that he did not care if the battalion could do communications or anything else, as long as this program was done correctly. It was instantly obvious that that one simple statement had shocked and demoralized everyone in the room. We were all dumbstruck by the fact that a communications battalion commander, or any commander, would stand in front of his people and tell them that he didn't care whether they could accomplish their unit's mission as long as it could do some other thing. It's even more baffling when you consider that the only purpose of the other thing was to support the unit in accomplishing its mission. That battalion commander lost a lot of people's respect and confidence in his ability to lead with that one hasty statement. It was clearly obvious he had lost sight of the purpose for him and all of us being there. By stating that the unit's purpose and mission was secondary to anything else, he had basically told everybody in the room that he was the leader of nothing.

Substance vs. Surface

Seek "earned" respect, not adoration. Adoration is fickle and fleeting. Don't try to be cool or liked. Be firm, fair, and consistent. People will follow you if they respect you, even if they don't like you; the opposite doesn't hold true. Adoration doesn't withstand trials and tribulations. Adoration is built on personality; personality is built on people's perception of you. The majority of people's perception of things center around how those things affect them. More specifically, their positive perception of things centers on whether the thing is beneficial or agreeable to them. If a leader's thinking or actions fall on the other side of the follower's perception coin, they will quickly find that adoration does not exist on that side. Leadership is too important to be left to people's varying perceptions.

Also, when you try to be anything other than who and what you are, people almost always know when you're being disingenuous. Discrepancies between who you are pretending to be and who you really are will eventually become evident to all. These discrepancies will lead to frustration and resentment for those who follow you because discrepancies equal inconsistency. Most people do not like fake people. If you are one, your followers may pretend to like you, but they really won't. Also, inconsistencies that result from trying to be who and what you aren't will eventually lead to actions on your part that will make you appear incompetent or unfair. So do not seek anything other than to be a good, principled leader and a good servant of the organization. When you aren't trying to be anything else other than who and what you are, people may not like it, but they will at least respect the fact that you are consistent and honest about your actions and intentions. Even if they don't like you, your followers will be able to adjust and become comfortable dealing with you if they know what to expect from you. But it's important to remember that being who and what you are in no way gives you license to be a butthole and to stay a butthole if you are indeed one.

Do the hard things.

Be willing, without hesitation or reservation, to do the hard things. Doing the hard things is what separates leaders from supervisors, managers, or bosses. The latter all too often go out of their way to avoid doing the hard things. Some just don't want to experience the discomfort or displeasure that often accompanies doing the hard things; others believe that avoidance of discomfort or displeasure is an inherent privilege that must be exercised. The hard things are what leaders and people in charge get paid to do. It's their only purpose for being there. If things never went wrong, if there were no issues to be dealt with, then organizations wouldn't put people in charge, thus saving millions of dollars in salaries. The unfortunate thing about being a leader is that you are infrequently made aware of the good things that are happening in your organization, but all of the bad stuff ends up at your desk. There is a term called *body hardening*, which basically boils down to applying repeated physical stress to a body part to "harden" it against future physical stress. This principle of hardening is not only applicable to the physical body, but to the mind and spirit. In the course of their leadership, a leader must have the self-discipline to engage in hard things that will harden their mind and spirit. This is what is meant by being willing to choose to do the hard things. The more they engage in them (hard things), the easier they will find them to handle and endure. Conversely, it becomes increasingly easier to slide down the slippery slope of choosing the easy way out the more you practice it. A leader must constantly guard against this path to becoming soft. A leader must be cognizant to build and maintain his leadership capability by choosing to do the hard things both in regard to others and himself.

Leaders have limits.

A leader has a finite amount of time, energy, and focus; they should use it judiciously and efficiently. This is true of all persons. In the end, a leader, even the best of them, is only human. If leadership is a good thing to provide to others, then we owe it to our followers to provide as much of it as possible, in as efficient a manner as possible. Optimizing our leadership means being judicious in the times, places, and ways we provide it. Whatever maximizes the positive effect it has and the number of people that it affects is the optimum provision of leadership. It is widely accepted that a single person can only realistically have an effective span of control of over five to eight persons, so it follows that the efficient provision of leadership entails the leader developing leaders amongst their subordinates and peers. Through this development, a leader is unconstrained in their leadership effect on the entire organization and its people. As a result, they shed the bounds of their personal limits.

Know a little about a lot.
Know a little bit about everything that concerns your organization/operation, and be an expert in the critical things that concern them. This is part of the science and the art of leadership. The science part comes in the form of knowing what the critical things are that you need to be an expert in and what things you only need to have a passing knowledge of. A good litmus test for a leader in determining what is critical or not critical is to determine what things, if they should fail, would cause the organization and/or leader to fail. Another way of

thinking about it is to envision glass balls and rubber balls. Glass balls will break if you drop them; whereas rubber balls, while you don't want to drop them, will bounce if you do. The goal is to not drop the glass balls, and if you do drop the rubber balls, don't let them bounce too many times.

At any one time in the organization, there are as many balls being juggled as there are things being handled. Since the leader is not handling anywhere near the majority of those responsibilities, that means his or her followers are the ones juggling most of the balls. One way a leader can try to ensure that those balls don't get dropped, particularly the glass ones, is by knowing what each one of those balls are, who are the people juggling them, and making it known to them that he always has an eye on their juggling act.

The art of knowing a little about a lot is realized by mastering the aforementioned two things to the point where it seems to your people that you know a lot about everything. To be truthful, the art part is really a psychological trick played on the follower by themself, and it is particularly effective the further removed the leader is from what the follower is into, whether that be in their work or private lives. Followers, on the whole, don't expect a leader to be interested in, concerned with, or relatable to the same stuff as they are. This expectation increases with the increase in position between the leader and follower. Some leaders reinforce this expectation because they see it as a safety buffer between themself and their followers; some enjoy it as an excuse not to know their people and look out for their welfare. However, those who enjoy and take advantage of the expectation create an opportunity for those who want to enhance their leadership. In his book the *Psychology of Persuasion*, Robert Cialdini talks about one the mental shortcut the mind uses to make sense of the world, which is to compare similar things with one another using the first thing experienced as a baseline. This comparison is called the Contrast Principle.[2] The thing about the Contrast Principle is that it, like many mental shortcuts, can be faulty at times. This is because it is a relative comparison and not an absolute one. Leaders need followers' buy-in to their leadership, and that requires trust. Trust requires belief that the person desiring the trust cares in some way about the person giving the trust. Showing interest in the things that another person cares about by possessing knowledge of it demonstrates caring. Leaders can take advantage of the Contrast Principle by knowing a little about a lot of different things where it concerns both the organization and their individual followers.

You are not the determiner of your leadership.
Never forget the led are the ones who determine who their leaders are. Leadership isn't bestowed upon anyone by a position or title, so if you do something to lose your

followers, you stop being a leader. Because of this, a leader must do their best to practice good leadership principles at every turn and avoid doing things that detract from their "leadership righteousness." Additionally, a leader must continually take the pulse of their people by whatever means they can to determine whether or not they are still leading them. The good news is that if you are doing a good job of leading people, indicators of that will present themselves to you of their own accord. The bad news is that if you don't check and continue to check periodically, you could lose your followers and not even know when or how it happened. Being in charge means people have to do what you say; it doesn't mean they have to follow you. There is a yawning gap between those two things. At the very least, the best efforts and results for your organization and its people will fall into that gap; at worst, failure will result because of it.

Leadership is one of those words that act as a noun in some instances and a verb in others. In the case of the leader who tries to provide leadership, the word is a verb. It's an action that the leader is trying to be successful at. In terms of the follower, the word is a noun. It is something that is given to them. But to be more precise, leadership as a noun is really something that the follower bestows upon the leader. It's something that they grant those they willingly follow. Followership is something that has to be willingly given to a person, thereby making them a leader of the person giving it, which is why it is truly the follower who determines who the leader is. Think of it as making coffee. When a leader assigns a task, they can't possibly spell out or even know each and every action required to accomplish that task. There are usually hundreds of actions required to accomplish even the simplest of tasks. Even if a leader is able to spell out 90 percent of those actions, that still leaves a lot unexplained. This is why the follower has a lot of power in the leader/led relationship. In actuality, the leader needs the follower to fill in all the other actions required to accomplish an assigned task. If the follower is incapable or unwilling to fill in those actions, they can follow the leader's orders and still have the leader fail. In the example of making coffee, if I tell a person to make a cup of coffee using the coffee machine in the office, that seems pretty detailed. But we all know there are many steps involved in making a cup of coffee, and if you break it down to the level of having to program a computer to do it, there are probably a few hundred steps involved in making it. The leader who gives the order presumes that the average person knows or can easily figure out how to make a cup of coffee and will do all the micro things to accomplish the macro thing of making a cup of coffee. However, that is in fact just a presumption. The danger for the leader is in either not recognizing that or believing it to be a foregone conclusion.

Be human.

A former company commander of mine was rotating to a new assignment, and we were having an outbrief (summarizing discussion when something has or is ending) as our leader/lead relationship was coming to an end. During command relationship outbriefs, what my commander and I always covered was telling each other what we believe the other did good, what could have been done better, and any suggestions for improvement. This particular commander, like most of the ones I have had, had done a very good job during his tour, and I told him as much. But I did comment on one thing I thought he could do better, which was letting people know that he was human. He was a little perplexed by that statement, but there were genuine and positive thoughts behind it. First, I thought he was a good guy, not only as a commander and Marine, but as a person in general. The problem was that he was extremely reserved and standoffish. Unless you spent some time interacting with him, you wouldn't be able to determine what type of person he was one way or the other. Now, some people will swear by this presentation of oneself as necessary component to being a good leader. However, I believe just the opposite. Yes, a leader should have a healthy personal separation from those whom he leads. As a matter of fact, I believe that it is required in order to lead effectively. But the fact of the matter is that leadership is a human endeavor. Only humans can truly lead other humans. It is why to this day, even with all the known imperfections of man, we still put humans, not machines, in charge of other humans. Quite simply, no one wants a robot or machine as a leader. Robots and machines can't and don't care. And if people don't think that the people in charge of them care about them, at least enough to not wantonly risk their life or welfare, they will never truly follow them. Another reason that a leader shouldn't present themself in a robotic or machine-like manner is that no one aspires to be more like a robot or machine. A leader who is hopefully setting a good example and doing what they believe is right wants their people to aspire to be like them. Another reason why a leader shouldn't present themselves in a robotic or machine-like manner is that there undoubtedly will come a time when the leader needs their people to simply trust their leadership based upon who they are as a person. Robotic supervision and machine-like efficiency will work fine for routine, non-stressful matters, but the time for personal trust of one's followers won't occur then. The time when personal trust is necessary for leadership will happen when time is critical and in short supply; when the consequences of action or inaction are dire; and when the opportunity to provide situational understanding to one's followers is absent. The tragedy related in the Mann Gulch fire comes to mind once again. The foreman, Dodge, had a very standoffish personality, and because of it, when the time for blind obedience to his leadership arose, there was no personal trust with which to garner unquestioned action from his subordinates.

Clean Slate

Try to start all new relationships with a clean slate, especially leader/lead ones. This is not to say that the wise leader does not try to get background on or gain insight into a new follower before they've developed a relationship. This is also not to say that the wise leader does not take under advisement what they've learned about a follower beforehand, whether good or bad. Quite conversely, having prior knowledge of a person's personality, competencies, and issues can be extremely valuable to the leader in laying the groundwork for the best chance at having a positive relationship with their new follower. However, the leader must take beforehand information and knowledge with the proverbial grain of salt as to its accuracy or legitimacy because these so-called facts can be skewed by the makeup of the outside source, whether human or machine. For example, you may be told by a former acquaintance of the person, that they are a jerk. However, the case may be that the two persons in question just didn't get along, or that the person giving you the information was the real jerk; therefore your soon-to-be follower is only a jerk in relation to that person. The leader must also recognize that past behavior is not an absolute predictor of future behavior. A follower's past behavior may have been decidedly affected by the situation they were in or where they were at in life at that time. Neither of which necessarily has anything to do with their new relationship with you. It is important to not let the due diligence of background research on a follower dictate the start of leader/lead relationship. This is important because for one, it's not fair to pre-judge someone. Second, through your pre-judging behavior, you could cause a potentially good relationship to be the self-fulfilling prophecy of a bad one. The simple fact is if you're going to have a bad relationship with a follower, then you're going to have it regardless of what you start out believing, so you might as well start out with the mindset that you're going to have a positive relationship. This makes even more sense when you consider that oftentimes people either live up or down to expectations. If you, the leader, have a negative expectation, then your perception of the follower's behavior could be skewed to the negative regardless of their actual actions. Or if the follower perceives that leader has a negative outlook toward them, they may then behave in an adversarial manner that would then be experienced by the leader as negative behavior.

Be selfless in carrying out your duties.
Selflessness as leader comes in two forms: sacrificial and outwardly beneficial. Most times when selfness is talked about in terms of leadership, it is defined around sacrifice and hardship on the leader's part. And while this is true and a significant component of selfless leadership, outwardly beneficial selflessness requires no sacrifice or

hardship at all. A leader can exhibit it simply by giving subordinates and peers credit where credit is due, even if it's a small thing and especially to superiors; looking for and taking advantage of every opportunity to praise and/or reward subordinates; actively seeking to pass on beneficial knowledge and skills to peers and subordinates; and not wasting the time or energy of your superiors, peers, or subordinates. The sacrificial form of selflessness is pretty simple in its makeup. If someone is left with a lack of, or if someone is overburdened, or if someone has to go the harder way or take the farther path, or there is pain or discomfort to be had, then you as the leader be the one to do it. If there is a choice between you sacrificing and others benefiting, or you not sacrificing and others not benefiting, or you not sacrificing and others having to, then you must choose the path of self-sacrifice. Selflessness, like other things, is contagious. Whether it be out of guilt or admiration, people will emulate you.

Constraint

Some believe that leaders are free to do whatever they want. Quite contrarily, they are frequently constrained in their actions more than their followers. This constraint is either imposed upon the leader by the rules, regulations, or policies of the organization, or the constraint is self-imposed as a matter of discipline. Make no mistake about it, constraint is a necessary discipline of leadership and therefore one of the hallmarks of a good leader. Having constraint can be broken down into two parts. The first part of constraint is knowing the purview, extent, and makeup of one's power and authority.

> **Purview**: Describes the areas, personnel, and things your power and authority covers. No one is omnipotent in everything.

> **Extent**: Describes the left and right lateral limits of your power and authority. This is distinctly different than purview. For instance, a unit's assigned personnel fall under the power and authority of the unit's commanding officer. However, while a military member's family is an integral part of their life, the commander does not have power and authority over the people in that member's family, so the commander's power and authority has its limits.

> **Makeup**: Describes your powers and authorities and, just as important, what mechanisms and in what ways those powers and authorities can be exercised. People often have the power and authority to accomplish the things they want to do but are either unaware of them or use the wrong one(s) for the situation.

The second part of constraint is related to the judicious, measured, and efficient use of one's powers and authorities.

Judicious: Refers to using your powers and authorities in a wise manner. How and when you use your powers and authorities are just as important as what powers and authorities you have. Power and authority alone do not solve or resolve every situation. Furthermore, the use of them in the wrong situation or in the wrong way can make things even worse.

Measured: Refers to exercising power and authority with the humbling knowledge of where your power and authority comes from. Very rarely do they originate from the individual who is exercising them. Exercise of power and authority should be done with the intended purpose always at the forefront of the leader's mind, and with the determination to utilize only the amount necessary to accomplish the purpose in an ethical manner.

Efficient: Refers to the powers and authorities granted to a person so they can carry out tasks. The person who is exercising them should be respectful enough of the granting authority to use them in a manner that optimizes accomplishment of the mission.

These three things make this part of constraint different from the first part because while you may have a large scope of power and authority, every situation does not require the maximum utilization of them, or the use of them in the same manner.

Lead from where you're at.

"Blossom where you are planted" is a phrase that applies to a person's leadership as well as other life situations. Leadership cannot be accomplished solely using positional authority. As a matter of fact, leadership should rely upon positional authority as little as possible. Many people believe they cannot lead without being in a certain position because their focus is on results. They want noticeable and big results so they can be recognized and lauded for their actions. And if that recognition can't happen, or if they don't believe they have a position that allows them to take actions that will later be positively recognized by others, they conclude they aren't in a position to lead. This is absolutely untrue because everyone is already leading and has been doing so since early in their life. The first person you ever lead is yourself, and from there, we have a myriad of leadership positions throughout life. We lead as siblings, parents, spouses, coworkers, and employees, to name a few. So accept that you have been leading (or failing to do so) whether you realize it or not. Leadership is all about development and growth. Development and growth prepare you to lead at the next higher level. So if you're waiting on being in a certain position to become a leader, then you probably don't have the competency or deserve to lead in that position. Also, a leader is who a person *is*, and leadership is what a leader *does*. Neither one of those things are dependent on something external like a job. Finally, a reason to focus on leading where you're at is that those who occupy their time and mind with other future leadership positions are likely to not be leading in the position they are currently in—or they are likely leading poorly.

Be a great you rather than a bad someone else.

Being a genuine leader is the only way to be a leader at all. You should leave your conflicting personal feelings, thoughts, and preferences out of providing the leadership that the organization desires. If those things are so conflicting that they are untenable for you to swallow or are outright deal-breakers, then you should lead yourself out of that organization. With that being said, you should also lead based on who you are, including your morals, values, ethics, and religious beliefs. In other words, give unto Caesar what's due Caesar, and give unto God what's due to God. In the book *Left of Bang: How the Marine Corps' Combat Hunter Program Can Save Your Life*, the authors relate that a principle of human behavior is that a person can only look natural doing something when they are doing something natural.[3] People can sense when others are being anything other than genuine, whether that be in a good or bad way. When a person's perception of someone doesn't dovetail with the actions and behavior of that person, it gets sensed as unnatural. Repeated unnatural behavior without explanation eventually leads to the belief that a person is being disingenuous, which is another way to say that they are socially lying to everyone. People don't trust people who lie to them, and without trust there is no leadership. Therefore people who are not genuine cannot lead others. Every individual is unique in their experiences, personality, skills, talents, and life situation, and anyone attempting to duplicate someone else will always do it badly. Conversely, being yourself is a natural thing and the easiest thing to do, right? Additionally, because you have a thorough understanding of yourself and have many practice repetitions at being yourself, and (hopefully) you recognize your potential and the need to be better, you can offer a great you to others!

CHAPTER FOUR: Detractors from Leadership

There is no such thing as a bad leader. The fact is that those people aren't really leaders at all!

There are just as many, if not more, pitfalls for a potential leader to fall into than there are places to ascend from. Many successful people haven't necessarily done everything right as they have simply not done anything catastrophically wrong. Unfortunately, not doing anything bad enough as to fall from grace does not mean that a person is a leader—not even those who reach very high positions. So how can the potential leader avoid the pitfalls that come with societal, professional, and personal success and advancement? Well, just the way that one avoids potholes in the road—by being vigilantly in looking out for them, recognizing one when they see it, and taking the necessary action to steer clear of them. The following covers some common pitfalls I have seen leaders fall into.

Believing Your Own Press

You can't afford to ever stop evaluating yourself and being self-critical. The higher you go, the less people there will be who are willing to tell you the truth. And the ones who are willing will do it less often—particularly if it is bad and particularly if it is about you. This is extremely dangerous for the leader and one of the biggest contributors to a leader's downfall. People telling you the truth as they see it, especially about yourself, is one external system of checks and balances against our selfish tendencies, which can lead us astray both morally and ethically. The reason it is so dangerous is because it creates blind spots for the leader. Blind spots for a leader, just like a driver, are rarely ever a good thing and pretty much always lead to bad events. Unfortunately, because of the tangible and intangible deference and benefits afforded to a leader, sometimes the leader is the only person to look at themselves with a critical eye. Therefore a leader must as act as their own final safeguard against their human fallacies. One of the best tools to combat this danger is humility. Someone once warned me, "You can either choose humility, or the Lord will humble you through humiliation."

Believing Might Makes Right

Just because you get to decide doesn't mean that you are right. Most times, a leader will be in a position to have the final say on a matter. However, a leader must never confuse having the final say with being right. If you are a leader, then you have followers. And good followers will provide opinion and advice, especially dissenting advice, right up until the time when the leader "slaps the table" on the issue—at which time they will faithfully execute the decision to the best of their ability. The important thing in this process for the leader to be cognizant of is that just because things work out (or at least don't go wrong) it doesn't necessarily mean their decision was right. This doesn't mean that a leader shouldn't use their best judgment to make and finalize a decision, nor does it absolve them of the responsibility to do just that. What it does mean is that a leader should be open to and consider all information and wise counsel when forming a decision. It also means a leader should guard against entering into or pretending to enter into the decision-making process while having already decided what to do. One of the silliest things I have seen leaders do is get people together to purportedly make a decision, then start off by telling everyone that they have already decided the matter and they are simply looking for anyone to put forth something that will change their mind. Look, if you have the power and authority to decide, and you've already taken it upon yourself to come to a decision by yourself, then just tell people what the decision is and get on with the execution. Doing anything else is an insult to everyone's intelligence and a waste of their time. People will know that you don't really plan on changing your mind and are being disingenuous. And if they don't know it the first time around, after you have knocked down any and all input and counter arguments in defense of your decision, they will figure out not to bother the next time. Worse, they will start to believe that you are a person who really doesn't want any information or input, especially input that is contrary to your opinion. This results in your people becoming yes-men. Just as often as not, things going right just means that nothing went wrong or the matter was one where the outcome of a certain decision made no difference one way or the other. A leader should always keep in mind the wise adage that says, "Two minds are better than one." The fact that only one mind ever makes the final decision doesn't change the advantage gained by having additional minds contributing. The other reason that might doesn't make right is that people eventually become resentful of and resistive to stuff being shoved down their throats. Even if the stuff is good and right for them, they will balk at the constant autocratic delivery of it. If a leader remembers that their "might" is not of their personal making, or for their personal gratification, or a de facto solution to getting every task accomplished, they will continually consult wise counsel, welcome sage advice, entertain dissenting opinion, and thoroughly deliberate on future decisions to give themself the best chance of making the right decision.

Pyrrhic Victories

Just because you get to decide doesn't mean that you've won. A leader should rarely have to "pull rank" in order to get their followers to do something. The leadership well will quickly run dry doing this. When I was serving in a position of leadership in the Marines, I used to tell my young Marines that the difference between someone doing what you tell them to do and someone following you is humongous. All rank means is that they have to do what you say. It doesn't mean that they have to follow you. I would tell them that when tough times come (and they will come), you give the proverbial shout "Come on, you sons of bitches! Do you want to live forever?", and you go over the trench wall, you'll either hear a thousand cries of your fellow Marines following you, or you'll hear crickets because no one is moving but you. In that moment, you will know with clarity whether you are the leader or not. The important part is that after you've gone over the trench wall is not the optimal time to find out one way or the other. Thinking you've won because you made the final decision on a matter is dangerous for a leader. For even the simplest of tasks, there are a thousand micro tasks that must be done in order to accomplish the primary task. Most times, a leader isn't able to fathom what all those micro tasks are—much less order that each one be done. A leader needs their followers to complete those micro tasks because they know that they need to be done to accomplish the macro task. In a nutshell, a follower can do exactly what they're told to do and still let a leader fail by simply not doing what they weren't told to do. This is what I mean by a pyrrhic victory. You can win the *telling* and still lose the *getting done*. If you are a leader, then you will be able to "sell" or "tell" your followers what to do and go for the full win because they will do it based upon a willingness to follow your leadership. That willingness will equate to them doing anything and everything they can think of to successfully accomplish the task without having to specifically be told to do those things.

Slipping into Hubris

Don't confuse position and authority with personal greatness and power. Leadership positions and the authority they carry exist to accomplish a mission. Only when it comes to leading oneself does the mission focus on the individual. All other leadership focuses on the mission of an organization or an external entity. This means that any leader who believes that their position makes them a great individual or that their authority is power endowed upon them to satisfy their own desires, is misguided at best and a contemptible person at worst. A person who needs a position to bolster their self-esteem or uses the authority of a position to lord over others is morally and ethically bankrupt as a human being and can't be an effective leader. The problem is that it is all too easy to let position and authority go to one's head. Especially if the person is not constantly looking for and guarding against the slide into hubris.

In 1993, authors Dean C. Ludwig and Clinton O. Longenecker published an article titled "The Bathsheba Syndrome: The ethical failure of successful leaders." The article detailed how many leaders are led to wrongdoing not because they are bad people, but because of their success. Success that has resulted from being good people or good at what they do. Their research found four snares of success that people are usually unprepared to deal with, which then leads them down the path to wrongdoing: 1) Success allows people to lose focus on the organization, or the mission, and develop complacency. 2) Success leads to privileged access to information, people, and things. These first two correlate to the positional hubris. The majority of workers are expected to produce and continue to produce some result that they are rewarded for in the form of pay. The person who thinks it is okay to no longer focus on producing the thing that they are paid to produce—because they have been successful in providing it before, or because the thing is now still produced even without the need for their input, or because no one has noticed or said anything when they stopped contributing to production—is stealing money! For the second of these two things, privileged access provides the means and opportunity parts of the means, motive, and opportunity triad. And since man's sinful nature always exists, there is a motive present as well. So you have a situation where the basis for all of man's actions are present, and the motive part is a negative one. Is there any wonder that what is done as a result will be negative? The final snares of success are: 3) With success comes more and more unconstrained control of the organization's resources, and 4) Success breeds belief in one's ability to manipulate outcomes. These final snares correlate to authoritative hubris.[1] The old saying is true: "Power tends to corrupts; absolute power corrupts absolutely."[2] Because success leads to increasing access to and unsupervised use of the organization's resources, success acts as a cancellation mechanism to an individual's external checks and balances. Unfortunately, for many human beings, having the authority to do something simply isn't gratifying without actually exercising that authority, whether it is legitimate to do so or not. The resources are the means to action; being unconstrained in their control of the resources provides the individual the opportunity to act. Finally, having had authority and having had success while wielding it can lead one to believe that they were the sole controller and determiner of the success that they achieved. This, in turn, can lead one to believe they have absolute and precise control over their success and that they have the ability to control and manipulate the outcome of everything. This is a fool's notion, as you might conclude. Eventually, somewhere along the line, their controlling action won't work as desired. Instead, the outcome will be unpredictable and undesirable. Nevertheless, the belief that one can always and perfectly control the outcome of things provides ample support to fuel one's motive for action.

Success: Fool's Gold

Don't confuse prior success with guaranteed future success. As often as not, past successes are merely failures that didn't occur, or decisions made by one that were ultimately proven successful by the individual efforts of many, or sometimes results that were made successful by other individuals in spite of a bad decision made by the leader. Because the leader often gets the credit for success, it is easy for the leader to start believing that success was solely or primarily due to their effort. As a result, the leader starts believing that everything they do is right and will lead to success. In other words, it is easy to develop the belief that one possesses the Midas Touch. Midas was a king who was so in love with gold he wished that everything he touched would turn to gold. He was granted his wish and soon discovered that his wish was a curse. Similarly, past success can be a curse on the leader's future. This is because past success is not a predictor of future success. So if a leader makes future decisions solely upon what was successful in the past and fails to address each situation as it is, they are not exerting a positive effect on the outcome at all. An applicable saying here is: "It is the fool who fights the war that he wants rather than the war that he has." A leader should only entertain himself with past successes in private and leisure, then put away his reminiscing of them as soon as that time is over. If one needs any reason at all for not wallowing in past success, they should only need to remind themself that their success is not theirs, nor of their own doing; it is that of the Lord and a testament to his goodness.

Even the Leader Is Only a Teammate

Leadership is a collaborative exercise. It is rarely done alone, at least not well. Good leadership is almost always the result of the efforts of multiple people. While there must be a head person in charge, that person's leadership is most often compounded and thereby maximized through the leadership efforts of others. A leader who fails to recognize this, or forgets this, is bound to have people fall through the cracks. This is especially true in large organizations or where complicated tasks are involved. In these types of situations, there are many people acting and multiple tasks are occurring outside the leader's immediate sphere of attention. Without the collaboration and cooperation of others in providing distributed leadership, there are bound to be parts of an organization or tasks that will be neglected. All this is to say that the leader is a position player, just like everyone else. If that was not the case and the leader could do everything, then there would be no other players, and the leader would be a leader of nobody except themselves. The reason why it is important for the leader to recognize that they are simply a position player is that everyone needs to play their position to the best of their ability in order for a team to be successful. The first step to playing a position well is understanding that you have a position. Then you must understand that the responsibilities of your position contribute to the success of the whole. After that, you will see how your actions in that position support that success. Without this clarity, it is easy for the leader to believe they are solely responsible for success. This faulty mindset may temp a leader to play another person's position, thereby hampering the success of the team and the individual.

Tongues Wag the Dog

Gossip is bad for business. A leader must quell the urge in themselves to gossip. Even if everyone else is doing it, gossiping is never a good look for someone in a leadership position. People will listen and even seem to enjoy it, but for the majority, gossiping will lower your standing in their mind, even if just subconsciously. Gossip is one of those things that makes a leader human in a bad way. Leaders should also avoid gossip because it rarely serves any purpose. If the leader is to expend their time and energy, it should be for something worthwhile to the organization or its people. Similarly, a leader must squash gossip from and amongst their subordinates because it sows distrust and promotes negative views about others (whether justified or not) that hampers people's ability to work together. It also detracts from the focus on mission accomplishment, and, by its very nature, it makes things personal that most often should remain professional.

CHAPTER FIVE: Followership

What the hell do those in charge know about leadership!

Supporting Leaders

Before you can hope to become a good leader, you must first become a good follower. And once you are a leader, you must continue to be a good follower. Why? Because everyone has a boss. Followership is the synergistic partner to leadership. It means doing and performing all manner of things there unto pertaining to supporting leadership and the accomplishment of organizational goals. So what goes into being a good follower? Again, some factors that support being a good follower can be identified.

Truthfulness

Truthfulness is at the core of being a good follower. The leader/lead relationship is bilateral and built on trust. If either side fails to uphold it, there is no relationship and thereby no leadership or followership. As a follower, being untruthful is one of the worst things you can do to a leader. Any leader with inaccurate info, whether it be accidental or purposeful, cannot make sound decisions. This generates blind spots and pitfalls for the leader. Dishonesty is the great sower of distrust. Therefore when the follower introduces dishonesty in the leader/lead relationship, they have now burdened their leader with the additional concerns of are they able to trust their people, how much, and whether they are getting their followers' faithful and best efforts? It should be a good follower's goal to never unnecessarily burden their leader with anything.

Providing a Sound Course of Action (COA)

While it is reasonable to expect that a follower will try to influence their leader's opinion and steer them toward their way of thinking, a follower cannot force their leader to think as they do and should not deceive them into it. Instead, a follower's duty is to provide their leader with viable COAs and as much decision-making info as possible to make a sound decision. If the follower's reasoning is sound and they do a competent job of presenting things, oftentimes the leader will arrive at the follower's desired COA all on their own. And if it is a matter of multiple COAs being equally viable, then it's incumbent upon the follower to present them as such and let whatever decision the leader makes be well with their soul. A quote from Colin Powell is applicable to this directive: "Avoid having your ego so close to your position that when your position falls, your ego goes with."[1]

Loyalty

Loyalty confirms that followership is a tricky thing. Many a good follower has gone astray by being loyal to leaders who have first gone astray. The tragic thing about this is that the follower was aware or suspected that the leader had gone astray but allowed themself to be led astray in the vein of remaining loyal. It is a follower's duty to point out to a leader when and where they believe they have gotten off track. If, after pointing out the error to the leader, the leader then continues to go astray, then the follower can only assume from their perspective that the leader is purposely going astray. And if that is the case, then the leader is no longer leading, and a follower has no duty to trail after them down the wrong path. However, being disloyal is not to be confused with a follower's difference of opinion with their leader. A follower's counsel to their leader should be the result of their best efforts to harness their knowledge, experience, and insight into sound advice. This means that sometimes the follower's opinions, views, thoughts, and feelings will differ from the leader's. The wise leader will welcome, appreciate, and utilize this counsel to come to the best decision that they can. In the case of minor differences of opinion, the follower has an absolute duty of loyalty. However, if the leader's opinions, views, thoughts, or feelings differ so much that they violate the follower's critical beliefs, then the follower should not be loyal; instead, he has a duty to resign from his position of counsel if he cannot change the leader's mind.

Furthermore, depending upon what real-world effects a leader's position may result in, the follower may have an even greater duty to attempt to curtail the leader's actions.

Play Backstop

In the end, a leader is only human. This means that they have as many frailties and are as fallible as any other person. Their weaknesses are simply different from others' weaknesses. A good follower will assess their leader's strengths and weaknesses. They will then encourage the leader in areas of their strengths and play backstop for them in the areas of their weaknesses. Part of a follower's playing of backstop concerns the organization and things external to the leader. As a backstop, the follower knows that the leader has a finite amount of time and energy with which to lead, and because of this, the follower will need to cover down on the other stuff that doesn't rise to the level of needing the leader's attention or require the utilization of the leader's time and energy to address. In the military, this is called "freeing up the commander to command." The other part of a follower playing backstop concerns the leader themselves as an individual. In this role, a follower will be a sounding board, provide a grounding influence, and act as an accurate and truthful mirror for the leader to see themself from the outside looking in. By playing back

stop in this manner, a follower helps protect the leader
the from their own frailties, fallibility, and sinful nature.

CHAPTER SIX: Relationships

"Marines are the Marine Corps."

Organizations are made up of relationships between people and tasks. The first of these relationships (between people) forms the basis for the other (between people and tasks). Relationships can further be categorized as formal, informal, task, or social. Regardless of how you classify them, cooperation amongst people is defined by relationships, and there must be cooperation in order to accomplish complex tasks. Relationships are simply either the result or the impetus for the aforementioned cooperation. When it's the former (the result of cooperation), the parties involved must give of themselves to make cooperation happen. When it's the latter (the impetus for cooperation), cooperation happens because people trust one another enough to do so.

Formal vs. Informal

One categorization of relationships is formal or informal. Formal relationships are those determined by the structure of the organization. For example: boss/employee, officer/enlisted, senior/junior. Formal relationships tie directly into how an organization's mission is accomplished and are therefore of utmost importance to the organization. Formal relationships are usually easy to identify, manage, and enforce. However, due to their critical contribution to mission accomplishment, extreme care must be taken in establishing logical and efficient formal relationships and monitoring/managing those that currently exist.

Informal relationships occur as a result of the interactions between human beings. For example: acquaintance versus stranger, senior employee versus new hire, likeable personalities versus distasteful personalities, dominant versus submissive, persons with similar jobs or interests, same gender and opposite gender. The important thing to note about personal relationships is that the informal ones can be just as powerful as the formal ones. The subconscious mind, in many cases, is more powerful than the conscious mind because we aren't aware of its workings. Contrast this with formal relationships and you'll see that a significant amount of the power comes from the fact that they aren't often recognized or thought about. Informal relationships tie directly into the welfare of the people in a group. That welfare determines whether they are willing to be part of the group and whether they are willing to do the work required of the group. Because of this, informal relationships must be of high concern to the leader. Unfortunately, informal relationships can be very problematic when you consider that they can easily become negative and detrimental, are hard to control, and can be impossible to repair. A leader must be as acutely aware of the informal relationships that exist as they are of the formal ones because on the negative side of things, the informal relationships have the potential to do much more damage than the formal ones.

Task vs. Social Relationships

Another important categorization is task versus social relationships. Task relationships can further be categorized as supporting/supported or complimentary/competing. In supporting/supported relationships, the successful accomplishment of one party's mission is dependent upon the other's support. In complimentary relationships, the accomplishment of one party's mission in conjunction with the accomplishment of a different party's mission serves in accomplishing a larger overall mission. Or the accomplishment of one party's mission simply enhances the accomplishment of a different party's mission. In competing relationships, the accomplishment of one party's mission comes at the detriment of the other party's mission accomplishment. Detriment in this case doesn't necessarily mean active harming; it could simply mean that one party's best chance for mission accomplishment is sacrificed to the other. The important thing about task relationships is that it is the most important relationship as far as a leader is concerned. This is because a leader's purpose is to ensure that the appropriate tasks are completed in order to accomplish the mission of the organization. While much of a leader's time and decision-making focuses on the social relationships, the only purpose for the existence of the social relationships is to facilitate task accomplishment. In other words, the creation of social relationships in an organization is the result of people being brought together to accomplish a task. Since the former (social relationships) are derived from and only occur because of the latter (task relationships), the leader cannot allow social relationships to hamper or become more important than task relationships.

The social type of relationship is usually defined by some combination of categories from the other three types of relationships. For example, a group usually consists of people who have social relationships as a result of having some kind of task relationship or mutual interests. At the same time, that group is usually tied to other groups or individuals through task or social relationships. The important thing for leaders to know about social relationships is that in any organization, relationships is one of the decisive factors in whether that organization is successful or fails in accomplishing its mission. In the book *The Five Dysfunctions of Team*, the author details how the dysfunction of *individual actors* causes an organization to fail, but he also explains how dysfunction between the *groups of individuals* causes the organization to fail as well.[1] The "so what?" for a leader is that if you want your organization to be successful at accomplishing anything, you'd better have a good grasp on the relationships in the organization even more so than you have of the technical aspects.

Opposite things usually attract (or complement) each other, compounding the other's effectiveness. Like things usually add no value at best, or repel at worst. A leader should avoid falling into the trap of surrounding himself with only like-minded or similarly suited individuals. This can be an easy trap to fall into because it's comfortable or convenient. What makes it even harder is that a leader is no different when it comes to their desire to be socially comfortable. However, where it concerns accomplishing a task, having people with differing viewpoints and competencies often goes much further toward maximizing the overall effectiveness of the group effort. Whereas having people with similar views and competencies most times only negligibly increases the group's effectiveness. This is because similarity adds no depth or breadth to the abilities of the group as a whole; it merely duplicates what is already there.

Task and social relationships correlate to the two concepts of task and social cohesion. Task cohesion and social cohesion are the two critical factors in getting anything accomplished that requires the efforts of multiple people. Notice that I use the word *cohesion* and not *harmony* or *homogeneity*. Nothing involving human beings will ever be any of those things, at least not perfectly. People are too varied in their individuality for this to be possible. Whereas words like harmony and homogeneity imply some perfect state of cooperation, cohesion merely indicates a sticking together of people, with no suggestion as to the internal dynamics of that stuck-together group.

Task cohesion is exactly that what the title implies: the combining of energies and the cooperation of efforts of two or more people to accomplish a task. There are only two instances where task cohesion is not important to people accomplishing a task. The first is if the thing is so unimportant that no one really cares if it doesn't get done. And if this is the case, no one should be wasting their energies or efforts on it. The second is if the task can be accomplished by one person alone. In which case, a leader has to ask themself why more than one person is involved or has been tasked?

Social cohesion is the willingness of people to interact with each other frequently and in a manner that will contribute to the successful accomplishment of the task that they have come together for.

There are two important things to know about task and social cohesion. The first is that both are required in some measure to successfully accomplish any task involving a group of people. The second is that the leader, who is entrusted with the responsibility of accomplishing the organization's mission, must ensure that social cohesion never supersedes or detracts from task cohesion. An organization gets together people to accomplish tasks that cannot be done by one person alone. So the people, from an organizational perspective, exist and are brought together for the task; the task does not exist for the people to be brought together. Jesus put it best when he said, "The Sabbath was made for man, not man for the Sabbath" (Mark 2:27).

The only reason you need social cohesion is to obtain task cohesion. Human beings are social animals. Cooperation is a social endeavor. That means if a task requires more than one person to accomplish it, some kind of socializing must occur between the people involved. Case in point: machines can work together all day, every day, to accomplish a task, and not a bit of social cohesion is necessary to make that happen. All that is required is that the machines be designed and set up in a manner that cohesively accomplishes the task. Letting social cohesion override task cohesion is the essence of putting the cart before the horse. A leader must tend to both, but task cohesion is primary.

Giving of Yourself

Because of their position and what they represent, a person in their role as a leader can have a much larger effect on people than they would if they were only judged upon who they were as an individual. The two greatest personal gifts a leader has that they can bestow upon a follower is their time and attention. However limited and fleeting these two things may be, a leader should never miss the opportunity to bestow them upon a follower. Leadership is a people business. Interaction between people is what it's all about. Because of this, not only does a leader bestow something positive upon a follower when they grant them their time and attention; the leader also receives positive feelings in return from the follower's response. The leader is in return bestowed with time and attention from the follower and has credits added to their leadership bank account by the follower because they are aware that the leader's time and attention is valuable; therefore they are appreciative when the leader spend time on them. Additionally, because human beings are naturally social animals, giving of oneself causes self-generation of a positive attitude and good feelings in the leader.

Trust

Trust is a critical component of leadership. A leader needs to trust that their follower will put forth their best effort to accomplish the mission. A follower needs to trust that the leader cares about them in some respect and knows what they are doing. Without trust, no leader/lead relationship exists. Trust can be generated in a number of ways. First, trust can develop naturally from the experiencing of trials and tribulations by a group of people. This is the underpinning of using hard training and unpleasant experiences, such as boot camp or fraternal pledging, to bond individuals together in a cohesive group. The dependency on one another that becomes necessary to survive and thrive in the aforementioned experiences causes the development of trust amongst those who go through it. In his book *Tribe*, Sebastian Junger talks about how the study of Native Americans during the formative years of the United States when they still roamed in tribes, found that they were generally happier and mentally healthier than European Americans because of their interconnected, communal living. Their way of living was based upon the fact that an individual couldn't survive alone, and everyone in the tribe was dependent upon each other to live.[2]

Second, trust can be generated by consistent behavior and demonstrating unselfishness. Consistent behavior equates to predictability. People are naturally comfortable with things they can predict. Comfortability with a person is one of the key ingredients of trust. Conversely, people are distrustful of those whom they believe are only concerned with their own self interests. One example of this is when a leader fails to stand up for what's right (be mindful to not misconstrue "standing up for" with "covering up for"), especially when and where it concerns their people and whether it works in their people's favor or not. When a leader knows in their head and heart what right behavior is, and they don't see it happening, at the very least they owe it to themself and to their followers to object on the matter. Even if the leader is told to shut up or is ignored, at least they will know they had the courage to stand up. More importantly, others will know, which will generate trust—even trust from the ones telling the leader to shut up and sit down. The simple fact of the matter is that cowards can't be leaders. A selfish leader violates the follower's need to believe that they are cared about. A leader who consistently demonstrates unselfishness develops trust in his motives from his followers.

Finally, uniform treatment of individuals is another way that a leader builds trust. Whether the treatment concerns discipline, reward, courtesy, or respect, the leader's treatment of followers needs to be seen as even-handed across the board. Additionally, when it comes to things like courtesy and respect, followers need to not only see they are being treated the same as all other followers; they need to see they are being treated the same as a leader's peers and superiors as well. Low is the person who is a kiss-ass to their boss and a slave driver to their subordinates. When people are not treated uniformly, someone will always believe that they are being treated unfairly in some way. People don't trust those whom they believe treat them or others with whom they identify, unfairly.

CHAPTER SEVEN: Communication

Move, shoot, COMMUNICATE.

From my experience and knowledge, the majority of problems and conflict between human beings stems from lack of communication, miscommunication, or improper communication. This is the reason why leaders should be extremely concerned with how they communicate and how communication is happening within their organization. However, problems and conflict are just the results of the higher-level reason a leader should be extremely concerned about communication, because communication is the primary mechanism for cooperation and collaboration between people in accomplishing any task.

Communication Roles

Communication consists of sending and receiving information. Therefore the two primary roles in communication are that of the sender and of the receiver. Each one of those roles has its responsibilities. The sender's responsibility is to ensure that the communication is successful. The sender is the one with the information that needs transmitting; therefore the sender is the one with the need to do something (otherwise known as a mission). The receiver's responsibilities are to be open and receptive to communication, including exercising active listening skills and being inquisitive in nature as far as trying to accurately understand what is being communicated.

Lack of Communication

The problem with lack of communication is that there is no such thing as a communication gap. If the sender fails to communicate a message to the receiver, then that in itself is a message. Also, humans tend to assign positive motives to their own actions and negative ones to others' actions. There is very little when it comes to human ongoings, where there is such a thing as a gap, because where there is a gap in something, someone will eventually fill it with something. This applies to communication as well. If a sender fails to communicate, then someone else (possibly the receiver) will fill in the gap in communication. And due to the aforementioned human tendency to think negatively where it concerns things external to themselves, the gap will probably be filled in with negative information for the sender.

Miscommunication

The crux of miscommunication is a difference in understanding between the sender and receiver as to what the message is. Miscommunication shares a similarity with lack of communication in that it facilitates the assignment of negative intentions to the sender on the part of the receiver. Miscommunication also hinders cooperation and collaboration because without a meeting of the minds, there can be no unity of effort in accomplishing a task. The other problem with miscommunication is not only can there be no unity of effort in getting a desired result, but miscommunication can lead to the generation of additional problems besides what was attempting to be addressed in the first place. Miscommunication is a primary disrupter of task cohesion.

Improper Communication

Improper communication (unlike lack of communication and miscommunication) doesn't involve whether a message was transmitted or if there is confusion on what the message was. The problem with improper communication is that the message, no matter how good, gets lost by how it is transmitted. Inappropriate communication turns off the receiver to the message, which means, just like miscommunication, there can be no unity of effort in accomplishing a task. Finally, inappropriate communication disrupts social cohesion like nothing else.

Overcommunication/Undercommunication

Where communication is concerned, both overcommunication and undercommunication are undesirable. Overcommunication wastes both the sender's and receiver's time by transmitting more information than is necessary. It also can cause the receiver to have information overload, which can then cause the receiver to miss or forget the important information that is trying to be relayed. Overcommunication can be exasperating to the receiver, resulting in them tuning out any and all communication from the sender.

Undercommunication leads to no understanding or a misunderstanding on the receiver's part. Obviously, without enough information, the receiver won't be able to establish an understanding of what is being transmitted. This means communication didn't occur at all. Similarly, a sender leaves the door wide open for miscommunication when they don't transmit enough information for the receiver to clearly understand the message being sent. In which case, once again, undercommunication equals no communication.

While neither over or undercommunication is desirable, if one must occur, it is far better to overcommunicate than to undercommunicate—because undercommunication equals no communication. Whereas with overcommunication, at least the message gets sent, which accomplishes something! And there is also the *chance* for it to be received successfully, even if it isn't done in an optimal manner. In other words, with the former (overcommunication) you run the risk of not effectively communicating, and with the latter (undercommunication) you guarantee it.

Poor Communication

There is very little excuse for unclear, unconcise, or inefficient communication. How well one communicates, for the most part, can be controlled. Attention must be paid to how important proper communication is to teamwork and people accomplishing a task. Proper communication requires the sender to focus on these four message transmittal techniques:

Grammar: Poor grammar hurts the ears—literally. A former English teacher of mine used to say, "A sentence is a sentence when it sounds like a sentence." Hurting someone's ears increases the chance that communication won't happen.

Sentence Structure: Most things can usually be said in several different ways by arranging words into a sentence. However, how the words are arranged can change the tone or the very meaning of what is being said.

Tone: Tone in itself transmits a message. What is important is whether that message is intentional or unintentional. Care must be taken to ensure the former and not the latter.

Body Language: The way our body moves can easily lead to poor communication when there is a message discrepancy between what is being said verbally and what is being communicate by our body.

The ability to effectively communicate can be developed and enhanced by the study and practice of professional communication, both oral and written. Communication skills, like anything else, get better with hard work and practice. Leadership is a professional endeavor, and communication is an essential part of it. As such, communication is one of the professional tools that are leader needs to master.

CHAPTER EIGHT: Self-Understanding

No one understands you better than you (you would hope).

Self-understanding is instrumental to a leader being able to lead effectively. The first person you lead is yourself. You cannot properly lead anyone without knowing and understanding your own motivations, desires, and weaknesses. Unfortunately, the easiest person to fool is also yourself; the easiest person to make excuses for or let slide is yourself; the hardest person to discipline is yourself; the hardest person to admonish is yourself; and the hardest person to hold accountable is yourself. I could go on and on, but the underlying point is that there are many natural human tendencies that fall on the negative side of the ledger when it comes to the never-ending endeavor to know and understand yourself. The best facilitators and educators in developing self-understanding are wisdom and maturity. By this I mean that wisdom and maturity come from self-understanding, and self-understanding leads to greater wisdom and maturity.

Self-Control

Don't be quick to emotion, especially anger. Strong emotions can get in the way of good leadership decisions and actions. The forefathers of the United Sates and the drafters of our Constitution purposely and wisely built in time delays to the process of making any significant change to our system of government. If you read the Federalist Papers, and I highly suggest you do, the reason for this was because the aforementioned drafters had a keen understanding of human nature, part of which is that humans are prone to quick and intense swings of emotion. As a result of these hasty and intense swings, humans are quick to form rash judgements and implement poorly thought-out actions. Furthermore, this tendency applies as much to the whole (group) as it does to the individual. When it comes to human decision-making and actions, the combination of emotions and thought, or feeling and thinking, are oftentimes a zero-sum equation. I used "and" rather than "or" because there is never a total absence of either one. However, using more of one leads to less of the other; hence the zero-sum designation.

My father used to say that a man who is angry can't think. This is why a leader should try not to be quick to emotion. Why? Because the more emotion is involved, the less thinking that occurs. I would even say that one should exercise *more* thinking than feeling, particularly for a leader. This is not to be confused with passion. Passion is warranted in certain situations, and extreme passion is called for in some. However, passion and emotions are two different things. There is an old saying that your first mind is usually the right mind, or your first thought is the best thought. You hear no such thing about your first emotion. Human beings love mental shortcuts because they make life easier to get through, so here is one for you: If there is a zero-sum battle between emotions and thought, you are more likely to get a good result if thought dominates over emotion. This is not to say that emotions are or will ever be totally eliminated from human action. It's just to note that one should not turn quickly to their emotions as the means of deciding or acting. Unless necessitated by the situation, decisions and actions should not be rushed.

The words *rash* or *hasty* as they apply to decision-making and action have rarely been a positive thing. Similarly, we should not be quick to judgement. The word *judgment* denotes that something is well-reasoned and well-thought-out. This is rarely the case with decisions and actions that are rushed. Judgement requires input, analysis, and consideration of pertinent information. Rarely is all pertinent information readily available, and thorough analysis does not lend itself to speed. Conversely, the shortcuts of bias and prejudice do lend themselves to speed. Unfortunately, they usually result in bad judgements.

Ego

Everybody has an ego. There is no need to deny it or minimize the effect it can have on your decisions and actions. Hopefully, you have developed a healthy super-ego to moderate both the id and the ego. There is nothing wrong with having a healthy ego; however, ego without the governing influence of the super-ego is bad. As a leader, an ego that is not governed and tempered almost always leads to inappropriate behavior and self-ruin. A good rule of thumb is that ego applied internally to one's discipline is usually good. Ego applied externally to others or the mission is usually bad. One of the previously mentioned leadership traits is unselfishness. The reason why ego applied externally is usually bad is because ego is usually self-centered. This self-centeredness and the resulting selfishness are directly opposed to the trait of unselfishness.

Ethics

Ethics is a concrete matter in some instances and an abstract one in others. As a leader rises to positions of greater power, authority, freedom of action, and freedom from oversight, the opportunities and potential to go off the proverbial ethical cliff become more numerous. Even good, well-intentioned, highly ethical people can be corrupted by ever-increasing means and opportunity to do wrong. All persons are fallible and corruptible through our sinful nature. Once again, the forefathers of the United States intimately understood this and constructed the entire framework of the U.S. Constitution around trying to ensure that these two things would not be able to override good and fair governance of men. The mechanism that they chose to try to prevent these parts of human nature from rearing their ugly head and to kill them if they did was checks and balances. This ingenious system has been wildly successful, especially considering it was designed hundreds of years ago. It seems to be in some jeopardy in modern times, but only because the three parts of our government that are supposed to maintain the checks and balances have increasingly shirked their roles in favor of other pursuits. However, the concept itself is as sound today as it was hundreds of years ago. And it is just as sound when applied to the individual as it is for a country. Everybody, especially leaders, needs checks and balances, both internal and external, if they want to be and stay a good person. Leaders face ethical pitfalls that make it even more important that they have personal checks and balances.

As a leader, ethics matter for these five key reasons:

- The higher position you attain, the less knowledgeable you are about the individual parts of your organization and the less aware you are of the day-to-day goings-on that keep the organization functioning.
- The higher you go up the less peers you have, which means less people to consult with and less people to keep you in check.
- The higher you go, the less you are forced to associate and work with people who have different views, opinions, and life situations.

- The higher you go up, the less people are willing to tell you the truth or bad news, or anything at all. This increases the likelihood that more stuff gets filtered before it gets to you (read "The Emperor's New Clothes").[1]

- The higher you go, the more responsibility you can shirk, the more access you have to resources, and the easier it is to believe that you are in control of everything (read "The Bathsheba Syndrome").[2]

Ethics can be an agreed-upon code of righteous conduct amongst a group of people, such as professional ethics, or it can be a settled-upon and adhered to code of righteous conduct by an individual. Either way, ethics are principles that are possessed and demonstrated through people's actions. The problem with ethics is that they are easy to have when there is little opportunity to be unethical or when a person isn't challenged. Having personal checks and balances, whether they be internal or external, supports the maintaining of ethics and informs what ethical behavior is in a given situation.

Ways Leaders Cultivate Unethical Behavior

They promote unmetered rivalry.

The promotion of "excellence as exclusion" and the cultivating of an "us against them" mentality happens amongst people in every way and manner imaginable. The purposeful use of these as tactics to promote healthy competition and/or encourage an attitude of excellence is the legitimate work of a leader. However, if not properly utilized and controlled, the "excellence as exclusion" and "us against them" mentalities can easily go wrong and turn into unethical behavior. All too often "us" becomes good and "them" becomes bad after "us" is determined to be better than "them". Or, the reverse happens, where "us" is deemed to be good and "them" bad first, which then automatically makes "us" better than "them." Whichever order it occurs, it usually doesn't take long after the "us" and "them" distinction is made for one or the other to transpire. It is animalistic human nature for people to want whatever group to which they belong to survive and thrive. This often has to occur at the expense of some other group not thriving or even surviving. For most individuals, their higher nature is not comfortable hurting or destroying other another individuals. One higher-level mechanism we have for getting around this is to create groups and then come up with a reason why the other group is less worthy of being treated how we would want to be treated. This is not an absolutely necessary component of the "us against them mindset," but it's something that often happens as a matter of human nature. The problem with making "them" bad is that it is the first step to making it okay for the "us" to do anything we want to the "them."

Not surprisingly, cultivating the "us against them" mentality is an easy way for bad leaders to hide and protect their own unethical behavior. By promoting an "us against them" mentality, the unethical leader gets people, both good and bad, to close ranks and insulate the leader and their bad behavior from outside forces that may be a danger to it. Another feature of this mentality is that the longer the "us against them" environment persists, or the higher the stakes of the competition, the more noble and righteous "us" becomes in the eyes' of "us," and the more sinister and eviler "them" becomes in those very same eyes. Case in point, I remember hearing a story about a youth sports team whose members were banned from their league because they had snuck into the locker room of an opposing team and defecated in the sport's equipment bags of the team members. Although young children had taken this action, there was adult involvement in the lead-up to the act. Now, those looking at such a thing from the outside would say that that is crazy. How could adults condone, much less encourage such a thing? But is it really that unbelievable or surprising? I can guarantee you that situation did not start off at that level. It almost certainly started off as a healthy competition and a good-natured sport rivalry. But with lack of control and active bad behavior on the part of the adults, the end result was something that even those very same adults probably can't now explain or justify.

They discourage communication and encourage secrecy.

Nowhere is the importance of communication or lack of it spoken to more powerfully than in the Bible where it tells of the Tower of Babel. What is told about the tower is that at a point in time all men decided to get together, cooperate and pool their efforts to build a tower all the way to the heavens. By cooperating and working together, they were actually able to accomplish the building of this tower to heaven. When God saw what man had been able to do, he destroyed the tower, scattered men across the face of the Earth, and gave them different languages so that they could never again get together, cooperate, and combine their efforts. By physically separating them and breaking their ability to easily communicate with each other, God prevented man from ever again achieving such extraordinary things through combined effort. Discouraging communication and encouraging secrecy creates what's called *compartmentalization*. Now, compartmentalization is great if you're trying to legitimately keep a secret, but in the case of the bad leader, if the thing that they're trying to keep secret is bad behavior, then compartmentalization is a tool for evil. Even if people see parts or individual actions of bad behavior, without having a full picture of what is happening or has happened, it is not likely that they will be able to truly comprehend what is occurring. And since people for the most part actively try to avoid conflict or difficulty, it is more likely than not that the person without knowing the full scope or consequence of another's wrongdoing will ignore bad behavior.

They remove oversight mechanisms.
Internal and external checks and balances form the locks to the gate of unethical behavior. Whether purposely or unintentionally, when a leader removes external checks and balances, they remove 50 percent of the unethical behavior safeties. Oversight mechanisms are one of the external checks and balances. When done purposely it's almost a guarantee that the leader has already begun behaving unethically, or definitely intends to.

They encourage unethical behavior by others.
This is a way that unethical leaders build a culture that is conducive to their own unethical behavior and how they insulate and protect themselves. By encouraging others to behave unethically, the unethical leader sets up a reverse extortion environment. If they get others to act unethically, then those people lose their ability to challenge the leader on his or her unethical behavior for fear of being challenged on their own; leaving those around the leader without a righteous leg to stand on. Additionally, and even worse, the unethical leader can use other people's unethical behavior as leverage against them in securing their support for the leader's unethical behavior. As far as building a culture, just like any

organism, unethical behavior can more easily exist and thrive in a similar and supportive environment. In this case, an unethical one.

Ways Leaders Cultivate Ethical Behavior

They use *please* and *thank you* indiscriminately.

A leader should communicate with subordinates on the same level of politeness as they would any peer or superior. This serves as a constant reminder to the leader that authority is to the position or rank that one possesses, not to the individual person. And the sole purpose for that authority is to accomplish the mission of the organization. Being polite to all persons keeps us from getting a "big head" and going off the rails. The leader who practices indiscriminate politeness is helped in remaining humble simply through the action of doing so. Politeness to peers and subordinates, even if only subconsciously, imparts that you have respect for them and the work that they do, understand that you need them as much as they need you, and that you look at the relationship between you both as one of teamwork and shared mutual goals.

They're cognizant of their authority limits, freely inform others what those are, and don't act outside of them.

For instance, when I'm dealing with a subordinate Marine or Trooper regarding a decision to be made or an action to be taken, I make it a point to tell them—whether it is me making the decision or taking the action, or whether it is someone else's authority to do so and I am merely acting as an agent of that person in either a directed or delegated capacity. Why is doing this so important? Because being cognizant of the limits of your authority and letting others know what those are, keep the leader from acting outside that authority. And acting outside their authority is what gets leaders in trouble. Being cognizant of the limits of your authority acts as a constant reminder that no matter how powerful you are, you still have limits, and limits keep you humble. Freely letting others know what those limits are is a way for you to put in place external checks and balances on yourself.

They explain themself.

As can be extracted from the Mann Gulch fire failed leadership scenario, there are multiple reasons why it is good for a leader to explain themself. One reason is so that followers can know and understand the leader's intended end-state. This allows those followers to be able to continue to work toward that end-state and theoretically success, even in the absence of further orders and when things go wrong. Another reason to explain oneself is so that followers can see that the leader thinks logically, has sound judgement, and is well-reasoned. Demonstrating these things before a crisis puts credits in the leader's leadership bank account that they can then draw upon to get followers' unquestioned obedience when there isn't time or opportunity to explain things. Another reason for a leader to explain themself is that it communicates to their followers that he or she sees them as important enough to make the effort to explain things. Whether real or not, this perceived caring leads to reciprocity on the followers' part. Besides all of these, another reason why it is important for a leader to explain themself is that a person having to explain their thinking and actions acts as both an internal and external system of checks and balances on that person.

They are self-employed and work alone.

No, this is not some morbid, antisocial manifesto. The concept simply recognizes that most people's problems when it comes to work involve the people they have to work with—whether they are superiors, peers, or subordinates. Or maybe it is more succinct to say that people make or allow others to be their problem. Work for yourself. You are the first person that you lead and should be the first person that you follow. In the end, even tasks that require a tremendous amount of people to get accomplished still boil down to individual efforts. No individual can be compelled to put in a good-faith work effort, so in the end, people really do work for themselves. They may be required to output certain work products for the employer and do so to certain standards, but to the extent that they do more or less within those boundaries is really up to their own personal motivation. When I say you should work for yourself, I'm not referring to owning your own business. What I'm talking about is a mental state and a way of viewing life. What I'm talking about is going about your work in a manner that when it's done or you are done, you are content with what was done and satisfied with how it was done. Working for yourself has many benefits. For one, if you fail, then you have no one to blame but yourself, which shortens the amount of time and energy you have to put into looking around for whose fault something is. If you work for yourself, then you determine what level of effort and results satisfies you, and as long as you have done your best to meet those expectations, you will be content. If you work for yourself, you won't desire or need external praise or recognition; your own satisfaction will be your praise. The easiest way to know that you are working for yourself is to perform your work at the same level and in as much the same manner as possible as if you were working alone.

They are introspective.
Self-analysis requires continuous intelligence gathering, processing, and evaluation, and it is a lifelong pursuit.

Here are some key questions a leader should repeatedly ask themselves are: "Do I demonstrate what I expect from others? Am I staying true to my morals, values, and ethics? Am I staying true to my organization's values, directives, and purpose?" A leader's continuous self-analysis is not confined to the strategic, big-picture level of life. It happens on the operational and tactical day-to-day level as well. A leader must be adept at making quick, introspective checks: "How much (energy and give-a-damn) have I got to give today? Where and how can I best apply *what I've got* to *what I have to do*? Is something really a problem and systemic, or is it just a one-off issue? Am I making a sound decision and taking well-reasoned action, or am I just doing something to do anything? Am I seeing and addressing the critical issues, or am I lost in the routine and inertia of day-to-day activities?" As the great philosophers Plato and Socrates believed, introspection leads to one's truths and is an important practice in human existence.

CHAPTER NINE: Actions

Words are meaningless, a man's actions are who he is!

A person is truly represented by their actions rather than their words, because actions require time, energy, and effort where words require very little, or none at all. I can say that I'm going to kick your ass, but unless I put the time and energy into doing so, the mere combining of words into a sentence that makes a statement means nothing. Because actions speak louder than words, one of the great burdens placed upon a leader is to try to not only talk the talk but to walk the walk as well. A leader must always be conscious of or at least believe that they are being observed and make sure that their words and actions never conflict. Because if they do, those observing him or her will choose whichever of the two is negative and assign that judgment to the leader.

A leader must try to measure their actions in terms of their appropriateness, intensity, and the resulting effects. A leader who fails to do so will come to be seen as wanton and callous. How a leader is perceived directly affects how effective they are. A leader must try to foresee how their actions will be perceived by others, not just what the leader's intent is in taking the action.

Decision-Making

Military strategist and United States Air Force Colonel John Boyd coined the term OODA Loop, a concept to help explain how to focus and direct one's energies. It's a cycle that stands for Observe, Orient, Decide, Act.[1] Decide comes before Act, but making a decision is an action itself. Many a leader has been tripped up by the inability, unwillingness, or failure to decide in a timely manner. Inability to decide, otherwise known as decision paralysis, can be caused by a number of different things, such as information overload (where there is just too much input for the brain to process or come to any reasonable conclusion about what action to take), fear of making the wrong decision, or the desire to garner more and more information. Fear of the consequences of being wrong can lead to non-decision by those with just plain ole lack of guts or character. Those who continually put off making a decision in hopes of receiving more information often do so thinking that with enough information, the decision will make itself, or the matter will simply resolve itself without having to have made a decision. Failure to decide in a timely manner can be caused by the person in charge not having enough knowledge or experience to recognize that there exists a situation that requires a decision (the "Observe" part of Boyd's Loop), or by them not being able to determine what needs to be decided upon (the "Orient" part of Boyd's Loop).

The overall point is that the first step in any action is the action of deciding what to do and deciding that you will do it. Before I go any further, I want to point out an important principle to the previously mentioned "Observe" and "Orient" parts of deciding. Many texts concerning decision-making assume that it is obvious and already known that a decision situation and point has been reached, and this is not the case. As I have already stated, this is one of the common reasons why leaders don't make a decision. When it comes to observing and orienting, the best assistant you have is the little man in your head. Everyone has probably heard the term that if it seems wrong, it probably is. While this adage is age-old, people ignoring it and running into trouble is as frequent today as it has been since the beginning of time. I imagine that Eve, upon first contact with Satan in the garden, suspected something was not right. In many instances, the subconscious mind is more powerful than the conscious mind; in other instances it's the opposite, and the subconscious is smarter and faster than the conscious. It is my contention that what we call the "little man in my head," or our "first mind," or "my gut," or "intuition" is merely our subconscious noticing something and processing the meaning and significance of it faster than our conscious mind. This bears itself out in that you, like me, have probably found that when you have the time and ease to think about a thing that was "bugging you," but you just couldn't figure it out, you often found the answer once you quieted your mind. Our subconscious is like a multicore computer processor versus the old single-core ones. It uses our present collection of knowledge, experiences, and animal instincts, along with current sensory input, to shortcut the process of "figuring things out." Because it uses these shortcuts, it can be fallible and therefore can't be relied upon as the sole or primary thing to inform decision-making. In his book *Blink*, Malcolm Gladwell talks about how the shortcuts that the mind utilizes can lead to wrong conclusions and bad decisions. However, because it can also be extremely accurate and fast, it is not something to be casually disregarded either—especially in critical situations where time is short supply and the stakes are high.[2]

Let me share a story from my law enforcement career. I was on routine patrol, just like any other day, when I was asked to patrol two towns up to the state border to BOLO (be on the lookout for) a felony suspect wanted for attempted murder. Law enforcement in the adjoining state had received a tip that the suspect may be passing through our area in the next couple of minutes. However, by the time our agency received the info and I was able to get to the area, it was long past the time of the tip, so I had little expectation of coming across the suspect. Of course, after having given up BOLOing and returning to regular stationary patrol, wouldn't you know that a vehicle somewhat matching the description of the suspect vehicle passed right in front of me. Even though I didn't believe it could possibly be the guy (because things only happen that way in the movies), I stopped the car and interviewed the driver. As I was talking with the driver, who was very polite and cooperative, the little man in my head began indicating to me that something wasn't right with him. However, his outward demeanor and behavior gave me no obvious reason to believe that anything was wrong, even after interviewing him thoroughly.

Short story is that it did end up being the guy, but the only way I knew that was that he suddenly and unexpectedly fled on foot from me. I chased him . . . yada, yada, yada. The point of the story is, I knew something was wrong with that guy after talking with him a while and before he fled, but I couldn't figure it out. Literally, after first talking to him, I went back to my vehicle and sat there a couple of minutes, asking myself what was it that was "bugging me" about him. And even when I was checking with our investigators to try to get a physical description of the BOLO suspect, I said to them this was probably not be the guy, but something just wasn't right about him. After the event was over, I ran it back in my mind, pondering what it was that kept telling me something wasn't right with that guy. And then *bang*, it came to me (the conscious mind, that is)! The entire time I was talking with the guy, he never once asked why I had stopped him! Of course, I had stopped him, asked him to step out of his vehicle, kept him there while I asked him specific questions about his goings and comings, separated him from his passenger and interviewed her away from him, then subsequently asked him to wait even longer while I went and checked some things. You know what? Never once did he ask, "Officer, why did you stop me?" I surely hadn't told him. So here's the point: my subconscious observed that small discrepancy between my knowledge and experience of normal human behavior in that situation; it observed long before the conscious mind could figure it out; and it used my knowledge of the situation (white male suspect fleeing and possibly passing through that area), my experience (this person hasn't done what almost every one of the hundreds of other people that I've stopped have done and what is reasonable to expect that a normal person would do), and the sensory input I was getting (the female passenger answered my questions but she was tense about something, and the vehicle generally matched the BOLO description), and it came to the correct orientation or conclusion in minutes.

This story is applicable to leadership in this way, if something seems wrong, then it probably is. If there are several somethings that seem wrong, it almost certainly is. This is important for a leader to know so that he does not summarily dismiss the little man when he starts knocking on the walls of his head. It's important to know because all too often, when a leader begins to make inquires, persons involved will give him the old, "You just don't understand!" or "We've always done it this way!" or "Nobody has said anything about it before." And those are just a few of many varied reasons people will give for things that are wrong. In these situations, a leader must continue to peel back the onion, ask the hard questions, and require concrete, legitimate, and independently supported answers. If a leader does not have at least some confidence in the abilities of little man in their head, they may miss throwing the B.S. flag when and where necessary and thereby miss an opportunity to eliminate one more possible pitfall for their organization and themselves.

Say what you mean and do what you say. This is critical for a leader on both the front end and back end of events. On the front end of things, people will judge your character, competency, and courage based upon whether or not you are true to your word. This will directly affect whether those people are willing to trust you, work for you, and follow you. Being a man of your word equals consistency and predictability. This gives comfort to others around you, even if the resulting consequences are negative for them. Predictability and consistency are two big things where it concerns the back end of things. As a leader, many ancillary events and actions of others will be the result of or based upon what you say or do. If people aren't able to know with some level of surety that what you say and what you do are valid measures of what to expect, then the result is chaos and disaster for the organization.

The *what* feeds the senses; the *why* informs decisions. It isn't enough to just know what; why is more important in all but immediate-action-required situations. Getting back to Boyd's OODA Loop, "what" constitutes the observe portion of the loop and "why" constitutes the orient and decide part of the loop. This is because once you orient yourself to what someone is doing, then you can decide whether or not what they're doing is in harmony or in conflict with what you're trying to do. At that point, you can decide what you're going to do about it. With a good amount of knowledge and experience, the why of what a person is doing will also allow you to reasonably predict what they will do in the future, which in turn will allow you to better decide and plan what action you're going to take.

Decision-Making Structure

Decision making is not an easy thing for the young leader, and it will never be an easy thing for the ethical and moral leader. Once you've come to a decision, I recommend going one step further if time and opportunity presents itself. In the space between deciding and carrying out that decision, I recommend applying a litmus test to your decision before executing it. A litmus test on the decisions you make is another one of those internal checks and balances that I talked about earlier that a leader should make regular use of in order to not go astray morally or professionally. Over the years I have developed a three-part litmus test that I apply to my decisions that has made decision-making easier, and I feel it has never led me astray. The first part is actually a four-part question of the decision: Is what I've decided illegal? Is it immoral? Is it unethical? Is it unreasonably unsafe? If the answer to any of these is yes, I have automatic cause for concern about the decision. The second part of the litmus test is a two-part questioning: Is it in keeping with the organization's rules, regulations, customs, traditions, and mission? Does it violate any societal, professional, or universal norms? If the decision violates any of these things, then it is almost certainly not being made in the best interests of the organization or its people and should be an immediate cause for concern. The last part of the litmus test is to question whether the decision I've made is about the organization and/or its people, or is it about me. In other words, who does it benefit? Oftentimes, a decision will benefit all three of those entities in some ratio. And there is nothing wrong with a decision benefiting the decision-maker, even if it's a simple thing, like getting something done the way you prefer. However, it becomes wrong if it's about me to the detriment of the organization or its people. Once again, the ego cannot be allowed to operate unchecked.

Decisiveness

An 80-percent solution enacted quickly and decisively is better than a 100-percent solution enacted too late. Once a leader has done their best to come to a decision on the best course of action, they should execute that course of action without further deliberation or timidity. There is no such thing as a perfect solution; therefore a person can never be 100 percent confident in any action. However, to not execute a well-decided action as if it is 100 percent guaranteed to work only invites the possibility of failure into the situation. There are an untold number of examples throughout history of failed actions that were great plans but failed in the execution phase due to half measures and lack of boldness.

Decisiveness cannot make up for poor decision-making or a bad plan. Execution of a bad decision or poor plan will merely lead to decisive failure. However, decisive execution as a result of good decision-making and proper planning can realize unimagined success. Decisive action can snatch victory from the jaws of defeat simply through force of will. The goal of combat is to inflict your will on the enemy. This can only be done through action. Decisive action, especially when coupled with inexperience, incompetence, or indecisiveness on the part of your opponent is a force multiplier of immeasurable value. Decisive action engages two of the tenets of MOOSEMUSS: maneuver and offense. Decisive action is maneuver because any action at all changes the situation from what it was previously by causing reactive events and actions. In other words, taking any action is a maneuver. Decisive action is offensive for the same reason; it causes things to react to the leader verses the leader reacting to them. Additionally, decisive action is offensive because the purpose is to force a situation's end result to be what the leader desires.

When taking decisive action, you should engage the other principles of MOOSEMUSS as well, such as:

Economy of Force: Economy of force does not mean you chintz on the amount of force used. As Colin Powell notes in his book *It Worked for Me*, during the Gulf War, when General Schwarzkopf asked him why he had sent him three aircraft carriers when he had only requested two, Powell's response was that it was his thinking that if Schwarzkopf thought he could get the job done with two carriers, then there was no question of the job getting done with three.[3] Economy of force simply means not utilizing more force or resources than is necessary for the action to be decisive.

Objective: If you're acting without having already established an objective, then there is no way that action can ever be decisive. What makes the action decisive is whether or not taking that action will give you a high probability of achieving your objective.

Security: Security where it concerns decisive action is merely holding your cards close to your vest. This does not mean that you husband information. You must ensure that those who need to know, do know, and they know everything that they need to know. However, if you let information about your thoughts or actions outside those people get out, your decisive actions become open to being foiled by others.

Unity of command: Unity of command involves not only designating who is in charge of who, but also ensuring that the command structure is set up in a logical, simplistic, and efficient manner that ensures the commander control of the resulting actions while allowing flexibility for those who need to carry out those actions. In assuring command and control, the structure of that command must be set up so that it guarantees that everyone is on the same page and is focused in the direction of the mission to be accomplished.

Surprise: Surprise is comprised of stealth and speed. Stealth requires hiding one's intentions and actions, or it can be achieved by the purposeful transmission of false intentions and action. Speed as surprise is simply a matter of taking action so quickly that you are far ahead of your opponent's ability to handle it with their OODA Loop process, or better yet, to even engage their OODA Loop. Surprise, while not a decisive action itself, is certainly a significant contributing factor to an action being decisive.

Simplicity: Keep it simple, stupid. Complexity invites disaster. Every additional bit of complexity added to anything you do is added risk of things going wrong. Simplicity helps those who are trying to execute decisive actions in multiple ways. It helps them understand the plan; it helps them remember the plan; and it helps them be able to adapt the plan when things go wrong while still accomplishing the plan's objectives.

Mass: Mass, or massing, is simply bringing together dispersed resources to generate a decisive amount of force at the appropriate place and time to achieve a decisive result.

One final key to decisive action is knowing where and when to take it. Decisive action taken in an ineffective manner or on an unimportant target is wasted. Decisive action not taken at the right time, whether it be too soon or too late, is useless because it will turn out to be not decisive at all.

Thinking vs. Feeling

Thinking and feeling, or your thoughts and feelings, are the impetus behind man's actions. To put it in other terms, your thoughts and feelings feed the motive part of the means, motive, and opportunity equation. Your actions are simply the result of your motives availing themselves of the means and opportunities that present themselves. Both thoughts and feelings will always be present in any action in some relative ratio. However, feelings are a fickle and fleeting thing, whereas thoughts, hopefully, are not. Where it involves actions to be taken, even if the initial impetus behind the action is one's feelings, I would strongly suggest that you apply a lot of thinking to deciding what action to take, including whether or not to take any action at all. Feelings serve us better as input to thinking and decision-making, rather than as the primary controller of actions. All animals have feelings. Even the most simple-minded animal can be frightened, amused, or angry. What separates us from them is a higher intelligence. So acting off of feelings makes us no better than a primal, reactive animal. Taking action almost always involves or affects other higher-level intelligence individuals or society in general. Because of this, a leader should do their best to ensure that their actions have a purposeful and desired effect.

Engaged Leadership

Inspect what you expect. When I say inspect, I don't necessarily mean the prearranged, clipboard and checklist-type inspection. Although, that type of inspection is a legitimate and useful tool that should be used in the overall concept of inspecting what you expect. Inspecting what you expect happens in both formal and informal ways, as well as directly and indirectly. For instance, the clipboard and checklist manner represent a formal way to inspect what you expect, but you can also accomplish it in an informal manner simply by walking around. Talk to your people and listen to what others are talking about—people at all levels and in various positions in your organization (that is, if you want to get a real feel and complete picture of what's going on in your organization). You can directly inspect people, processes, and things, or you can do indirect inspections by observing indicators of the state of those things. Things like your people's pride in their appearance, the material condition of equipment and gear, disciplinary issues, as well as positive recognitions, retention numbers, customs, and courtesies, or how people generally treat each other—all indirect inspection points. The great thing about informal and indirect inspections is that they can be done more easily and more often than the other types. Also, more times than not, they provide the leader with a more accurate and insightful gauge of what is going on because they provide for an unvarnished observation and evaluation.

So what does "inspect what you expect," entail? First, be where no one expects you to be, and show up in the places no one wants you to go. If people are doing the right things and doing them in the right ways, there should be no reason why anyone, including the boss, can't show up and be present. If righteous things are occurring, then those things should occur in the same manner, whether you're there or not. If the way business is being done changes when you show up, then something is wrong. Either things aren't being done right when you're not there, or things are aren't being done right when you are there. Both of these situations are bad because theoretically your expectation is that things are being done the right way (according to legitimately established standards and procedures) all the time.

Second, ask the questions people don't want to answer. If people are doing the right thing in the right way, they should know and understand what they are doing. If they know and understand what they are doing, they should be able to answer most questions about what they are doing. Just as important, they should be happy and more than willing to answer questions about it. If your people can't answer reasonable questions about what they're doing or are hesitant to do so, then you have a problem. This means you either have people out there acting without proper direction, or you have people out there being given improper direction and they know it.

Third, check out and check on the things no one wants you to pay attention to. When people are not doing right or purposely doing wrong, the last thing they want is the leader's attention. Those not doing right will go to great lengths not to do anything that will cause the leader's attention to turn their way. Those purposely doing wrong will do everything to keep the leader's attention turned away from them. You can address the former by proactively spot-checking things within your purview. As for the latter, the leader has to develop a keen sense for when they are being steered, maintain an overall interest and cognizance of all aspects of their organization's operations, and cultivate an environment where when they see something wrong, people at any level, in any position, are willing to raise their hand and keep it held up until you see it.

Tamed Bull in a China Shop

Sometimes people get comfortable with the way things are or have always been done. In other words, they start considering their environment, their current world concept/construct to be a china shop—unassailable and unerring. This can happen even with things that those very same people dislike or believe are incorrect. Simply as a result of familiarity, people will adopt and stringently defend things that are objectionable to them. Therefore in order to get people to change, grow, or innovate, you have put yourself in the role of a bull in their china shop. Albeit a tamed one, but a bull nevertheless. One who consciously and purposely breaks china, but only when appropriate, and only some or certain china; a bull who breaks china with an eye toward either clearing some space so that shop shines more, or getting rid of junk to make room for real treasures. A leader should stretch their people by pushing their limits and moving them out their comfort zone. Muscles only get stronger when stressed, and people's habits are the same. Many of mankind's greatest achievements and advancements have come in times of stress or distress. To cite a famous quote, "Necessity is the mother of invention." There is no fun or growth in doing what you already know and have proven that you can do. Conversely, there is degradation from stagnation and atrophy of the will when people are allowed to rest on their laurels. Challenging people keeps them interested and engaged. It gives them a reason to get up in the morning, provides them with "good for the soul" work, and gives them something to accomplish that will result in a feeling of self-satisfaction. The simple fact of the matter is that great people do hard things, and hard things forge great people.

CHAPTER TEN: Caring

Caring is the root of all good leadership.

The greatest leadership trait and principle that I've learned over the years is that of caring. I say that it's a leadership trait and principle at the same time because caring is a personality characteristic, and it is also an actionable item. Caring also comes in the internal and external forms. My father latched on to an old Hallmark advertising slogan many decades ago and used it repeatedly throughout the time he was alive. The slogan was, "When you care enough to send the very best." This remains in my head till this very day not only because he used it so much, but because it speaks to internal caring. The slogan points to "you," the person who initiates and does the caring. The slogan says "care enough," indicating that there can be varying types of care. Finally, the slogan says "to send the very best," stating that a person has control over how much they care and how much effort they put into caring. When it comes to leadership, this doesn't necessarily mean sending a Hallmark card, but it does mean doing all the things that would be best for the people that you lead.

Caring as a Trait

A trait is commonly defined as "a distinguishing quality." Rather than being scientific, consideration of whether caring is a trait is empirically based. The fact is that people care about some things and not about others. The two primary purposes of leadership are to accomplish the organization's mission and to take care of its people in doing so. The focus of the former purpose is the organization, and the latter is its personnel. Keeping this in mind, a person could either care or not care about either the organization or its people. Most everyone will agree that it is quite evident from a person's behavior when they do or don't care about something. Most people would also agree that there are recognizable good or bad effects, both internally and externally, on a person as a result of whether they care. Therefore caring can thus be qualified as a distinguishing quality. Caring as a trait is made up of many of the leadership traits outlined in chapter two. For example, if one cares about the people they lead, they will be *unselfish* in how they lead, they will be *loyal* toward the organization and its people, and they will perform their work with *enthusiasm*.

Caring as a Principle

The word *caring* is defined in such terms as "being concerned about," "having regard or thought for," and "exhibiting kindness or empathy." Notice that being, having, and exhibiting are all action words. A principle is a rule of thumb from which to base one's actions. Each one of the previously mentioned leadership principles requires a proactive effort or action on the leader's part. Because of this, caring is a contributory factor, if not the direct motivation, for each and every one of those principles. You have to first care in order to want to set the example, know yourself and seek self-improvement, or know your people and look out for their welfare. A person who doesn't care will take no action or make any effort.

Caring Internally

Internal caring derives from a person's character. Once again borrowing from my father, he used to say, "Try your best to be the best at whatever you do. Even if it's being a floor sweeper. You try to be the best floor sweeper there is." What he was talking about was the internal type of caring. Some people call it work ethic; some call it having pride in your work, but it's all the same thing. It is the willingness and effort in expending our most precious assets, time and energy. Only things that we care about do we as humans beings expend either one of those things on.

Caring Externally

External caring is the manifestation of a leader's internal caring. It means praising those we lead when they deserve praise, encouraging them when they need encouragement, and disciplining them when they need correction. It means being unselfish even if you don't want to. It means showing enthusiasm even when you don't feel it. It means doing what's right and best rather than what's popular. It means going the extra mile even though you were exhausted ten miles ago. It means doing what needs to be done without being asked, without being known, and without ever being recognized. It means doing the right thing even when you know that people will be ungrateful, suspicious, or downright hostile. It means all these things and more if you believe in your heart and mind that they are the best things for the people whom you lead.

CHAPTER ELEVEN: Professionalism

Be the best floor sweeper there is.

Competency

Competency can only be demonstrated. It is the only way to prove that one possesses it. Just as the led determine who is a leader, whether you're competent or not is determined by those external to yourself. Competency can be demonstrated in two different ways: through examination or action. Examination requires one to prove competency through either interrogation of one's grasp of the subject matter, or through one's voluntary discourse on it, whether that be written or verbal. The second way that competency can be proven is through one's actions. Through exceptional execution of tasks and/or fluency in utilizing the knowledge, skills, and abilities associated with the tasks, a person proves their competency.

Competency of the Leader

A leader can prove his or her competency in the following ways:

Through examination. Written tests, interview boards, professional research and publishing are some easy examples of formal examination of competency. Besides being subjected to formal exam, speaking or writing intelligently on a matter is one of the easiest ways to demonstrate competency. Writing, in particular, allows you to broaden the population of people to whom you can demonstrate your competence because it doesn't require one's physical presence, it eliminates any possible detractions from the communication as a result of how it is presented by the individual, and it can be distributed widely and rapidly. Advances in modern technology with video, streaming, podcasts, etc. has in most ways imparted the same benefits to speaking as writing. However, there is still the possibility of message interference due to the presenter's presentation. Speaking and writing also has additional benefits as far as proving competency. Professional speaking or writing kills two birds with one stone in that when done well, a person proves their competency in the area of communication in addition to their competency in whatever they are speaking or writing about. Closely related to professionally speaking and writing is engaging in spirited discourse with others who are themselves competent on a matter. Besides showing one's own competence on the matter, discourse has the added benefits of introducing one to new ideas and viewpoints. Good discourse helps a person to identify weaknesses or inconsistencies in their thinking and develop a holistic understanding. Discourse also strengthens a person's competency by having to defend their ideas and viewpoints. Good discourse helps develop a person's familiarity and fluency with a subject through repetitive engagement, helps develop one's powers of persuasion, and helps develop professional and personal relationships with like-minded and opposite-minded people.

Through action. Being known to competently do and have done the thing(s) in question. Actions truly do speak louder than words. This is nowhere truer than when it comes to proving one's competency. Repeated performance of a task with a high degree of success in accomplishing it is proof positive of competency in and of itself. While accomplishing the task itself is important, just as important to proving your competency is explaining the thinking and reasoning behind your actions to those whom you wish to believe in your competency. As a leader, this also kills multiple birds with one stone. First, even if people see you successfully accomplish a task, if it is a mystery to them how or why you are doing it, they are just as likely to believe it to be anything other than competence. Second, explanation helps develop future leaders, builds followers' confidence in your leadership for those times when you don't have the time to explain, and gives the commander's intent so that followers can accomplish what you want in the absence of further direction.

Competency of the Follower

A big part of leadership is about sowing seeds, the growth of which benefit the follower, the leader, and the organization. The other side of the competency coin for a leader is developing it in their followers. The process of leading others is to teach, train, and task, then provide the necessary authority, resources, and support, and get out of the way. As you can see, the first part is of the process is to teach, train, and task. These three things fall into the domain of competency. A leader should not task a person with anything before ensuring that they have the competency to accomplish it; unless of course they are willing to unnecessarily risk failure. Either way, tasking a person with something before ensuring that they have the competency to accomplish it is called setting a person up for failure. A basic level of competency can be determined by ascertaining whether a person has the knowledge, skills, and abilities (KSAs) to accomplish a task. A leader can't reasonably hold someone responsible for failure to accomplish a task if the follower did not possess the KSAs to do so. If a leader finds that a person does not have the KSAs to accomplish a task, then it is incumbent upon the leader to teach or train them before tasking them, or not task them at all. However, if a leader has done their due diligence in trying to provide the follower with the necessary KSAs and the follower still can't or won't grasp them, then the leader should get rid of that follower. Or, as the great strategist Sun Tzu contended, "If his instructions and words are not clear and thoroughly understood, then the general is to blame; if the soldiers are not properly trained and equipped to obey his commands then the general is to blame; but, if his commands are understood and the soldiers are properly trained and equipped and they still fail to obey, then it is their immediate leader's fault and that leader's head should be chopped off."[1]

Another part to developing a follower's competency is that the leader must demonstrate confidence in it once they believe the follower possesses it. People do actually live up to standards and expectations. Showing faith in a person's competency will, in many instances, cause them to set high standards and expectations of themselves in order to be worthy of the faith shown in them. People must be put in a position to succeed or fail at a task based upon their own judgement and actions before they can truly be considered competent. In the process, mistakes and lessons learned will increase their competency level.

Study Leadership

If you are reading this book, then you have at least begun one of the most important duties that a leader has, and that is the study of leadership. Why is it the duty of a leader to study leadership? Well, it is my contention that the answer to the age-old question of whether leaders are born or made is that leaders aren't born. The good news is that because leaders aren't born, then pretty much anyone who is willing to dedicate themselves to becoming a leader can become one. I say that leaders are not born because the intricacies of leadership are too vast to be possessed simply through birth. One can possess many of the attributes of a leader and have a good foundational start in life in being provided the things that go toward being a good leader, such as good ethics, morals, and values. But possessing those things does not guarantee one will become a leader. There is an irreplaceable need for experience in dealing with people. There is the need to develop a unique leadership style that works for the individual, and that can only be fleshed out over a time and through trial and error. The common denominator of all these things is that they require time and work in the field of leadership in order to be developed. Since no one is born with time or experience, leaders are not born. This leads us back to why it is the duty of a leader to study leadership. The study of leadership only enhances one's leadership abilities, just like study in any other field or profession. At the very least, one is enhanced by study, even if they only learn what not to do.

Have a Repertoire of Styles

Being a SEA has some unique aspects to it, one of which is that as often as not, you are assuming a very important leadership role in a unit where you have little or no technical knowledge of the unit's primary Military Occupational Specialties. So as a leader you instantly lose your expert authority, at least in the area of the unit's mission essential tasks. Since the reason for a unit's existence is for it to be able to accomplish its mission essential tasks, the loss of this type of authority is significant. Sometimes you get a unit where your mission as a leader is to simply maintain the course. Sometimes you get a unit that just needs a little course correction, and sometimes you get a unit that needs a full 180-degree course change. Regardless, you must gain the trust, respect, compliance, and cooperation of your subordinates, peers, and leaders alike. This is where having a repertoire of leadership styles comes into play. There are both organizational and people reasons why you must have competency with different styles. On the organizational side of things, you clearly can't expect to lead your peers and superiors in the same manner as your subordinates. So the structure itself dictates that you have to utilize different leadership styles. Where it concerns people, each individual is infinitely unique and therefore most effectively led in different ways. The larger repertoire of styles you have to choose from, then the more likely you are to have the most effective one available to apply, which allows you to reach a wider range of people to whom you can provide optimally effective leadership.

CHAPTER TWELVE: People

I only know two things about people. One is, I don't know anything about people, and two, it's all about the people stupid.

I end with the last chapter being about the most important thing in all of leadership: the PEOPLE! Whether the people represent the leader or the people they are trying to provide leadership to, leadership's purpose, focus, center of gravity, and success or failure is dependent upon what does or doesn't occur with the people involved. How do I know that people are the most important thing in leadership? Simple! Without people, nothing gets accomplished. Leadership doesn't get accomplished, followership doesn't get accomplished, and the mission doesn't get accomplished. The second way that I know that people are the most important part of leadership is that without people, leadership means nothing. Without people, there is no one to be led. Without people, there is no one to do the leading. Without people, there is no purpose to leadership.

The leader that understands the following will have already gone a long way to being successful at leadership. A leader needs to realize that their greatest effect on people, and thereby the organization, is not necessarily tied to their personal actions. For most things, the people external to the leader have the most direct, immediate, and observable impact on mission accomplishment and are the center of gravity around which leadership revolves. This may sound elementary, right? You may say "Of course!", but this foundational leadership concept is lost on many a would-be leader. And it is no surprise that this is the case. Immediate events, as well as other people's issues and the leader's own issues, can easily obscure that everything stems from people, involves people, and revolves around people. If you are a leader and you have a leadership problem that seems incomprehensible, unsurmountable, or is just scrambling your brain in general, then I would encourage you to step back from it to gain a new perspective. Then focus on understanding that people somewhere in the situation are the cause of whatever is happening. Try to discern what person or persons are having an effect on the issue and what those effects are. More importantly, try to discern why they are purposefully or inadvertently having an effect of the issue. The leader who develops a solid grasp of the aforementioned things in any situation can almost always determine some way in which to alter the situation in order to move toward a desired outcome. At the very least, receiving clarity on the issue will help relieve the frustration you encounter.

If a person can master the people, they can master leadership. There are two categories of people that must be mastered in order to master leadership: oneself, then others.

Mastering Oneself

If you hope to successfully master the people you lead, you must first master yourself—or at least start on the path. Mastering oneself consists of disciplining oneself in emotion, thought, and behavior. In the universe, energy is neither created nor destroyed; it is simply transferred or transformed. When energy is transferred to another form in a suboptimal exchange, there exists what is called *waste energy*. Similarly, in leadership, waste energy is representative of when the leader's own "negative stuff" gets in the way of their optimum leadership efforts. To counteract waste energy, you must implement discipline, have perspective, and learn the art of mastering yourself.

Discipline

Disciplining oneself as far as the willful and purposeful control of one's emotions, thoughts, and behavior is the key to mastery. Mastery over these three things will best allow the leader to eliminate them from the equation as much as possible when leading and in dealing with leadership issues. One way to begin disciplining oneself is to reflect on how emotions and thought patterns have affected your behavior in the past. Another path to discipline is to study how others' emotions and thoughts have affected their behavior. Then examine what impact that had on their situations. Whether it's through reading, interviewing, discussions, or personal observations, make maximum use of learning vicariously through others' experiences. Finally, another method of developing personal discipline is being cognizant of times and situations where your thoughts, emotions, and behaviors are impinging on leadership issues. From there, you can consciously practice exercising discipline over your emotions, thoughts, and behaviors as you encounter new leadership opportunities.

Perspective

Mastering oneself also includes learning and knowing oneself not only in a first-person perspective, but from the second-person (reflective/critical/analytical) and third-person (objective) perspective as well. We as human beings likely (or at least hopefully) know the most about ourselves from the first-person perspective, often spending a great deal of time concerned about and trying to affect how we are seen from the third-person perspective. Ironically, we pay little attention to who we are from the second-person perspective. Funny how these perspectives mirror the whole id, ego, and super-ego setup, huh? Knowing ourselves from the first-person perspective is the key to understanding ourselves, bettering ourselves, and having positive regard for ourselves. Knowing ourselves from the third-person perspective is key to achieving the effect we want to have on our lives and the people in it, which includes those we lead. Being able to reliably see ourselves as the rest of the world sees and experiences us allows us to take action or adjust our actions in a manner that provides the best chance of successfully achieving our goals. It's imperative to examine the second-person perspective, the one that we have the least grasp on and concern for, because it's the one most conducive in maintaining positive internal checks and balances, keeping us on the path of righteousness, and contributing to us being a good leader. The reflection, analysis, evaluation, and criticism of the second-person perspective keeps us humble and constantly seeking righteous perfection.

Mastering Others

If you master the *internal* (you), then you will clearly understand how to master the *external* (others). Mastering others has nothing to do with lording over them. On the contrary, it means mastering the observation and discernment of personalities, traits, and tendencies. It requires mastering the art of engaging with others in order to progress toward a goal's accomplishment. It requires mastering both the science and art of taking the various experiences, backgrounds, teachings, motivations, and talents of those whom you work with and harnessing it all in a coherent and useful manner, then focusing it on the accomplishment of a goal. As you can see, mastering others requires work on the part of the leader rather than the leader benefiting from taking advantage of others' work. The path to mastering others travels through one's own thirst for knowledge and growth, humility, proper priorities, and life balance.

Thirst

The thirst for knowledge and growth is the only reason to try and be a good leader and continue to develop as one. Without it, you are probably one of those people who thinks they are already optimally effective (and likely you're the only one who believes it). Since you have reached that milestone, then there is nothing else for you to do. You probably believe that everyone should recognize you as being a good leader—and those who don't are either fools who don't understand your methods or are simply against you. Thirst comes from never being fully satiated and from having a desire for more. You can always obtain more knowledge, sharpen your skills, and develop new abilities. In doing so, you will experience increased personal and professional growth.

Humility

Humility directs a leader's focus away from I as the center of gravity and toward the people. With this correct orientation, a leader's thirst for knowledge and growth will naturally take them in the direction of mastering the other things necessary in order to manage and lead people effectively. Without this, a leader is almost certain to fall victim to the human tendencies of self-centeredness, obstinance, and selfishness.

Priorities

Balance and priorities are key to making sure that your pursuit of leadership excellence is lifelong and sustainable. There are three key areas where your priorities matter the most as you learn to master yourself: God, family, and people, in that order.

My recommended first priority is a relationship with God. All things flow from and fall naturally into place when we have faith in God, trust in him, and remain steadfast in our belief. The Lord walks with us through good and bad—whether we do good or bad. Without him as the highest and ultimate priority, everything else we do, particularly where it concerns others, will just be a hodgepodge of inconsistent and inconsequential things of the moment. With him, all things are connected with a greater and all-encompassing purpose. They happen for an intentional purpose and for our good. A belief in him and a faithful walk in how he desires us to live will leave you never bewildered, at a loss as to what to do, or without the strength to do it. This is the greatness you perceive in most great leaders whether they profess it or not. A greatness not of themselves but of the Lord that shines through them. When it comes to God as your first priority, a leader need simply keep in mind that nothing you are engaged with or can do on this earth is more important than or worth violating his direction and teachings for how we are to live. If you never take your eye off of the beacon of goodness and righteousness that is him, then it will be impossible to spiral into wrongdoing and ruin.

My recommended second priority is family. Organizations, jobs, positions, and purposes can all be great, but they all come and go. Most will be what they are whether you ever existed or not and none will ever love you. In his book *Detroit: An American Autopsy*, Charlie LeDuff said something I found extremely profound. He said that a family's job is to be there and see you through from birth till you get put in the ground.[1] No person or pursuit in life will ever be happy to see you just because it's you—besides family. No one will ever call you Mommy or Daddy like it is the most awesome thing in the world—besides your kids. No one will ever tell you that they love you and make you feel like you would move the world to make them happy—besides your spouse. Family can and will do all that stuff with nothing asked or expected in return. So take care of your family. And when I say *family*, I mean you the leader as well. There is a saying in law enforcement when it comes to emergency driving: "If you don't get there, nobody gets helped." This means if you drive in such a manner that gets you into an accident on the way to something, then whatever it was that required a police officer's response never gets one. Worse yet, now additional resources will be required to respond to your accident besides those that have to respond to the original incident you were going to. When it comes to taking care of yourself, the same concept applies. If the leader isn't healthy and operating at peak performance, then leadership doesn't get done for your followers and, worse yet, eventually someone will have to tend to you when you go down.

My recommended third priority, along with the organization, is to the people. The people are the ones who do the work and get the work done; therefore they are the organization, and accomplishing the mission is wholly dependent upon them. In most cases, taking care of one (the people) is taking care of the other (the organization). If you don't take care of one, the other will surely suffer, or at least not operate optimally.

Balance

The key to a leader maintaining a healthy balance in all aspects of their life is a healthy perspective of life in general and leadership in particular. Five concepts of a healthy leadership perspective I've observed are:

Leadership occurs everywhere. We lead in many areas of life that we don't consider a leadership position, but if you're a husband, father, mentor, or coach, I guarantee that you are leading people. You're leading whether you know it or not and whether that leadership is positive or not. The point is that you shouldn't get wrapped around the axle about any one of these many leadership positions, or your successes or failures in them. You can be succeeding in one while failing in the other, so success or failure doesn't hinge on any one position or decision. Your only litmus test should be if you are satisfied with your efforts, based on time and calculating the totality of your achievements.

Everything may not get done. We all have our idea and vision of what the perfect situation and/or outcome would be. This includes all the wonderful things we would like to do and accomplish as a leader. However, the truth of the matter is that there is only one of you, and you only have so much time and energy in this life, which means there is a limit to how much you are going to get done in it. The important thing is to know which things are *glass balls*, which things are *rubber balls*, which things are *silly putty*, and which things are *chewing gum*. Glass balls will break when dropped and cause you great consternation as a result of their breaking and being unable to be replaced. If you let your marriage fail or your kids end up hating you because of your work, then you've dropped a glass ball. Rubber balls are things that aren't desirable to drop and that you shouldn't allow to bounce more than necessary. But if you do drop them, the result won't be catastrophic to you. Silly putty represents the things that are fun to interact with and are nice to have, but if you drop or put them down and happen to leave them behind, it is no big deal either way. Chewing gum represents the things that once you have chewed them and they no longer possess any sweetness and you spit them out on the sidewalk, you should never pick them up and put them back in your mouth.

Most things that occur aren't significant, even things that seem world-shaking at a moment in time. My father had a saying, "A hundred years from now it won't make any difference." Well, I've found that with most stuff, a hundred seconds from now it doesn't make any difference. Forget about the amount of stuff that will be insignificant a hundred minutes, hours, weeks, or months from now. I can almost guarantee if you stopped right now, looked back at the times people offended you (if you can remember), you wouldn't be able to conjure up anywhere near the same level of emotion now. You likely won't even remember what the offense was even if you remember the person who offended you. Or it's just as likely that you won't even remember the person or offense, just that at some point in time you were offended over something.

Even if you do a good job and have an impact, you probably won't get to see the results. This throws many people off kilter because we need to see the results of our work to feel validated as a good worker. The results are equivalent to indirect praise. Recognize this need for validation for what it is. It is simply a desire to have the ego petted. Once recognized as such, since it is a desire and not a necessity, you can let go of that desire when you feel angst because you don't or won't get to see the results of your work. If it is good work and you were faithful in doing it, then you should know and trust that somewhere, at some time, there will be a good result—whether you know about it or not. In other words, even if you don't hear the tree fall in the forest, it still fell and somebody did or will hear it, or see it on the ground in time.

Don't expect praise or gratitude. Both require putting your happiness and contentment in the hands of others. If you don't expect praise or gratitude, and aren't looking to get them from others, then your happiness and contentment resides within you. If those things reside within you, then in essence you work for yourself. Knowing that you work for yourself frees you up to work for others, and true leadership is all about working for others.

CONCLUSION

So what, if anything, can be taken away from what I have laid pen to paper for (or fingers to keys as is the modern-day case, LOL)? Well, while I tried not to overwhelm you with my religious beliefs, I hope I made it clear that religious faith is the basis for true achievement in life. Being successful at leadership is definitely an achievement, and I assume most of you are in some process of getting there or are improving where you are at. Leadership is an achievement of faith and is directly tied to the role faith plays in the leader's life. The two primary components of leadership are the PEOPLE and CARING. Both of these things are only dealt with successfully from a perspective of faith, and all other things involving leadership flow from these two elements. So I wrap the topic with God first, then family, then occupation. Leadership is required in all three of these things, but with God, you should always be the follower and not the leader.

Another takeaway that I hope resonates with you is that you (the leader) are your worst enemy and best ally. The flaws and frailties of an individual are too numerous to list. However, they are only dwarfed by man's capacity to do good and achieve unimaginable things. The only thing standing between the bad and good part of man's nature is the greatest gift that God has bestowed upon us besides our time and energy: our free will. Our ability to choose is derived there from. And while such a choice is not a one-time, all-encompassing, unchanging thing, you as the leader can and should make the purposeful decision to "go forth and do great things" in as many ways, places, and as much as humanly possible.

The final takeaway is that leadership is complex in some ways but simple in most (treat others how you would like to be treated). It is a science through processes and procedures and an art through judgement, intuition, and perspective. It is definitive in some respects because it takes a person to lead other persons. And it's abstract as well because you can achieve the same results in different ways utilizing different combinations of styles and methods. It's unpredictable because you can be leading one minute and following the next, or doing both at the same time. Whichever way leadership is presenting itself, you should be giving it your best effort because everything you do matters and people are always watching and looking to you, hoping to find a respectable, honest, outwardly focused, and empathic leader to place their trust in.

Notes

Chapter One: Sound Principles

1. Sinek, Simon. *Start with Why: How Great Leaders Inspire Everyone to Take Action.* London: Portfolio Penguin, 2019.
2. Theodore Roosevelt quote, BrainyQuote®, https://www.brainyquote.com/quotes/theodore _roosevelt_140484.

Chapter Two: Exemplary Traits

1. Maslow's Hierarchy of Needs, Britannica, https://www.britannica.com/biography/Abraha m-H-Maslow.
2. Mann Gulch fire, Wikipedia, https://en.wikipedia.org/wiki/Mann_Gulch_fire .
3. Enter, Jack E. *Challenging the Law Enforcement Organization: Proactive Leadership Strategies.* Dacula, GA: Narrow Road Press, 2006.
4. *Wyatt Earp.* Film. Warner Bros., 1994.
5. Martin Luther King Jr. quote, BrainyQuote®, https://www.brainyquote.com/quotes/martin_lu ther_king_jr_109228.
6. *U-571.* Film. Universal pictures, 2000.
7. Freud, Sigmund, James Strachey, and Anna Freud. 1999. *The standard edition of the complete*

psychological works of Sigmund Freud. London: Hogarth Press and the Institute of Psycho-Analysis.

8. Henry Ford quote, BrainyQuote®, https://www.brainyquote.com/quotes/henry_ford_145978.

Chapter Three: Contributors to Leadership

1. *Pure Country*. Film. Warner Bros., 1992.
2. Cialdini, Robert B. *Influence: The Psychology of Persuasion: Robert B. Cialdini* (New York, NY: Collins, 2007).
3. Horne, Van Patrick, and Jason A. Riley. *Left of Bang: How the Marine Corps' Combat Hunter Program Can Save Your Life* (Old Saybrook, CT: Black Irish Entertainment LLC, 2014).

Chapter Four: Detractors to Leadership

1. Ludwig, Dean C., and Clinton O. Longenecker. "The Bathsheba Syndrome: The Ethical Failure of Successful Leaders." *Journal of Business Ethics* 12, no. 4 (1993): 265–73. https://doi.org/10.1007/bf01666530.
2. John Dalberg-Acton quote. Member, Acton Staff. "Lord Acton Quote Archive." Acton Institute, July 9, 2022.

https://www.acton.org/research/lord-acton-quote-archive.

Chapter Five: Followership

1. Colin Powell quote, AllAuthor, https://allauthor.com/quotes/127899/.

Chapter Six: Relationships

1. Lencioni, Patrick. *The Five Dysfunctions of a Team*. San Francisco, CA: Jossey-Bass, 2012.
2. Junger, Sebastian. *Tribe: On Homecoming and Belonging*. New York, NY: Twelve, 2016.

Chapter Eight: Self-Understanding

1. Hans Christian Anderson, "The Emperor's New Clothes" *Fairy Tales Told for Children*, (C. A. Reitzel, 1837).
2. Ludwig, Dean C., "The Bathsheba Syndrome," 265–73.

Chapter Nine: Actions

1. Luft, Alastair. "The OODA Loop and the Half-Beat." *The Strategy Bridge*. The Strategy Bridge, March 18, 2020. https://thestrategybridge.org/the-bridge/2020/3/17/the-ooda-loop-and-the-half-beat.

2. Gladwell, Malcolm. *Blink: The Power of Thinking Without Thinking*. New York, NY: Back Bay Books, 2019.

3. Powell, Colin L., and Tony Koltz. *It Worked for Me: In Life and Leadership*. New York, NY: Harper Perennial, 2014.

Chapter Eleven: Professionalism

1. Sun Tzu quote, Quotepark.com. https://quotepark.com/quotes/2087468-sun-tzu-if-words-of-command-are-not-clear-and-distinct-if/.

Chapter Twelve: People

1. LeDuff, Charlie. *Detroit: An American Autopsy*. Westminster, London: Penguin Books, 2014.

About the Author

William James Singleton was born in Manhattan, New York, and raised in the Bronx. He graduated from Alfred E. Smith Vocational High School and joined the United States Marine Corps Reserve in June 1995. He attended recruit training at Parris Island, South Carolina, and after graduation attended Marine Combat Training and Cryogenics military occupational school. William served at five different units, including four as the unit's Senior Enlisted Advisor. William retired as a Sergeant Major from the Marine Corps Reserve in 2022, after twenty-seven years of service, holding the final billet of Group Sergeant Major for an aircraft group comprised of ten separate units and approximately 3,000 personnel. William received numerous awards and decorations during his service, including the Legion of Merit.

In 2004, William joined the New York City Police Department. He graduated as the Honor Sergeant of his recruit class and served as a patrol officer in the Bronx until 2006.

In May of 2006, William joined the New York State Police. He graduated at the top of his recruit class and subsequently served as a road Trooper until 2014. Afterward, he served as a Sergeant for several years at multiple locations and currently holds the rank of Lieutenant.

William holds a Bachelor of Science Degree from the State University of New York: Empire State College and a Master's Degree in Public Administration from Marist College in New York. He is an active member of his church.

William is married to his wife, Racquel, and has three children, William II, Gideon, and Genesis.